Building on a Budget: A Practical Guide to Cost Management for South African Construction Projects

Nicholas

First Printing, 2024

TABLE OF CONTENTS

PAGE NO.

TABLE OF CONTENTS (Continued) PAGE NO.

TABLE OF CONTENTS (Continued) PAGE NO.

TABLE OF CONTENTS (Continued) PAGE NO.

TABLE OF CONTENTS (Continued) PAGE NO.

CHAPTER ONE : INTRODUCTION

CHAPTER ONE
INTRODUCTION

1.1 Introduction

Cost planning and control is possibly one of the most important functions that the quantity surveyor performs (Tan, 1988). Much of this planning and controlling process involves the preparation of price forecasts at various stages of the design. These price forecasts serve to predict the probable price of, and thus the client's likely financial commitment to, that design.

Ashworth (1988) and Seeley (1983), amongst others, have highlighted the importance of cost planning and control. Due to the prevailing economic climate in South Africa, where there is limited funding available for construction combined with high inflation and interest rates, this control process is of critical importance. However, cost control can only be effective if the estimator is able to forecast the tender price with "reasonable accuracy".

"Reasonable accuracy" is a relative term, being entirely dependent on each individual situation. What is important, is not that the required level of accuracy is achieved, but rather that the estimate fulfils its function, which is to ensure that the cost control is effective in all its objectives.

1.2 The Problem

Estimating is not an exact science, but, by definition, is subject to uncertainty. It involves the prediction of a future event, which in itself can never be irrefutably achieved since the future is not known and the present is constantly changing. Thus absolute accuracy in price forecasting is impossible to consistently attain.

Absolute accuracy is seldom expected. However, "reasonable accuracy" in predicting the accepted tender, is. Such a level of accuracy is essential to facilitate the cost control process. Morrison (1984), Beeston (1987) and Tan (1988), amongst others, contend that the actual levels of accuracy achieved are inadequate, and thus the cost control process is unlikely to satisfy all its objectives.

No such studies have been conducted in South Africa and thus quantity surveyors are unaware of the level of accuracy that they attain. An awareness of their forecasting accuracy and the factors which influence it will contribute towards enhanced performance.

1.3 Hypothesis

It is hypothesized that, by means of an analysis of the accuracy of a sample of forecasts, it is possible to identify the most influential factors which contribute to the attainment of the accuracy by quantity surveyors in South Africa.

1.4 Objectives

The objectives of this dissertation are:

1. To determine the actual levels of price forecasting accuracy achieved at tender stage and to compare these with the levels perceived to be attained by professional quantity surveyors. This will indicate whether quantity surveyors are in fact aware of the degree of accuracy which they achieve.

2. Thereafter, by analysing the relationship between the price forecasts and lowest tenders, errors and trends can be established to ascertain which factors affect the forecast's accuracy.

1.5 Methodology

A literature search was conducted to review the publications relating to the accuracy of price forecasts. To substantiate the problem, data was collected to measure the actual accuracy levels of price forecasts in practice. An opinion survey was then conducted to determine the accuracy levels that quantity surveyors perceive they attain in their own estimates and the factors which affect their accuracy. From the data of measured accuracy levels, the factors which affect accuracy were determined and conclusions were drawn.

CHAPTER TWO : THE THEORY OF ACCURACY

CHAPTER TWO

THE THEORY OF ACCURACY

2.1 What is Accuracy?

Accuracy may be defined as the "nearness to the truth" or the exactness with which an estimated value represents the actual value (Stevens, 1983, p.13). Ogunlana (1989) and Flanagan and Norman (1983) simply describe it as "the absence of error". Accuracy in the context of this thesis is a measurement of how closely the quantity surveyor's forecast can predict the accepted tender figure.

2.2 Error, Bias, Consistency and Accuracy

By measuring the level of accuracy of a forecast one is simply making an assessment of the degree of error that it contains. Since both the quantity surveyor's forecast and the contractor's estimate are subject to error, neither will represent the actual or "true cost" to construct the project. The term "error" in forecasting is thus relative, as there are no correct or "true costs" from which it can unequivocally be determined - only variable parameters against which it may be compared. Therefore, even if the quantity surveyor were to estimate the "true cost" of the project, this estimate may not necessarily be accurate. This is due to the fact that the contractor's estimate will almost always contain a degree of error, and any forecast which does not contain the same degree of error will appear inaccurate when compared against it. Errors arise due to many factors which predominantly emanate from the uncertainty inherent in the forecasting process. These errors may be either constant or variable. Constant errors are defined (Stevens, 1983) as the difference between the average of a large series of measurements (i.e. the forecasts) and the expected value (i.e. the tender values). The measurement of constant errors determines the degree of bias. A very low bias, that is a small or insignificant average difference between the forecast tender figure and the actual tender figure, is often associated with a high degree of accuracy. This is not strictly true.

Variable errors, measured by the level of consistency, are defined by Skitmore (1990) as "the degree of variation around the average". Thus the lower the degree of variation, the greater the consistency. Consistency is sometimes equated with precision (Stevens, 1983; Ogunlana, 1989) and efficiency (Skitmore, 1990). The measurement of these errors is discussed in detail further on in the chapter. The following four graphs on the following page illustrate the concepts of accuracy and consistency.

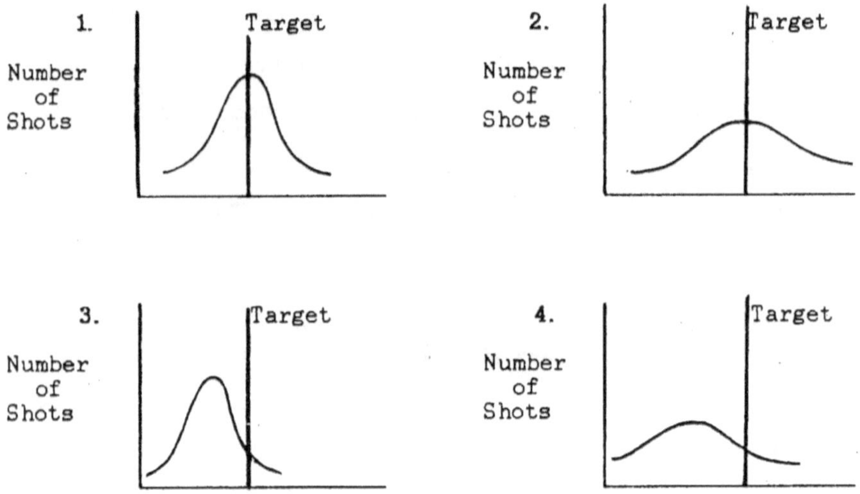

FIGURE 2.1 THE RELATIONSHIP BETWEEN ACCURACY AND CONSISTENCY (Adapted from Ogunlana, 1989)

The above graphs depict the principles of accuracy and precision by using an analogy of four riflemen each shooting at a target. Graph 1 shows the ideal, where accuracy is high and the shots fired are consistent. Graph 2 shows a good accuracy but the shots have a high degree of variability and thus exhibit no constant error. Graph 3 shows a high consistency but low accuracy, i.e. a high uniform error. Graph 4 shows both low accuracy and precision.

Since accuracy is an assessment of the degree of error, both constant and variable contained in the forecast, it is thus a measure of the combination of bias and consistency into a single quantity.

2.3 The Importance of Accuracy

Cost planning and control is possibly the most essential task that the quantity surveyor performs. The importance of the cost control function has been attributed to the following factors (Ashworth, 1988; and Seeley, 1983) :

1. The prevailing economic climate where there is limited capital available for construction, combined with high interest and inflation rates, makes an accurate prediction of price of vital importance.

2. The greater urgency for the completion of buildings. Clients do not have the time to allow for delays caused by redesign when tenders exceed the forecast price.

3. The requirements of clients are becoming more complicated and more consultants are being engaged in the design process, which makes effective cost control not only more difficult but also more important.

4. The larger client organisations (many of which use sophisticated forecasting techniques themselves) expect a high level of efficiency and expertise in controlling the cost.

5. The use of new and innovative construction techniques,

materials and methods of design makes the assessment of costs more complex and opportunities for substantial cost overruns are greater.

The above explains why the cost control of the design has become an essential task performed by the quantity surveyor. Since price forecasting is the key activity in this control process (Property Services Agency, 1980) the quality of the estimates will to a large extent determine the effectiveness of this control process. These estimates must be sufficiently accurate and reliable to enable cost control to fulfil its objectives, which are (Ashworth, 1988) :

1. To give the client good value for money.
2. To achieve a balanced distribution of expenditure throughout the elements of the building.
3. To determine the expected expenditure and keep it within this amount or the amount allowed by the client.

Thus the importance of accuracy in price forecasting lies in the forecast's function as the fundamental component in the cost control and planning process.

2.4 The Consequences of Inaccuracy

The preliminary estimate which the quantity surveyor prepares at the inception stage of the design usually establishes the cost limit. This cost limit is then used to ascertain the feasibility of the project which will determine if the contract is to be carried out. The accuracy of this initial estimate is thus crucial to all parties concerned. An overestimation of cost may result in a viable project not going ahead. An underestimation, however, could result in an unrealistically low cost limit being set which will cause complications when more detailed estimates are prepared or when the tenders are received.

Where the cost is inaccurately forecast and the tenders are greatly in excess of the figure predicted, the client has three options :

1. to abort the entire project and accept the losses from the expenditure already incurred.

2. to continue with the project as planned, but to obtain additional finance.

3. to change the design by reducing the scope and/or quality of the project.

The graph below illustrates how an inaccurate forecast (both an under-and an overestimation) will result in greater expenditure for the client.

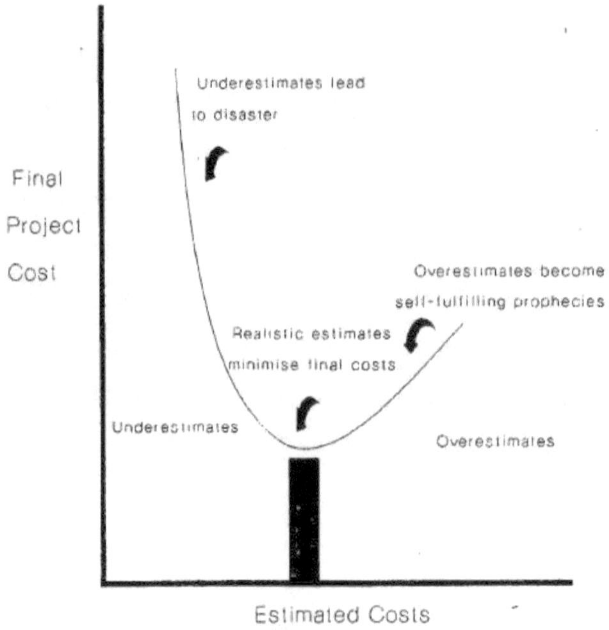

FIGURE 2.2 THE FREIMAN CURVE
(Daschbar and Agpar (1988), cited by Ogunlana (1989))

2.5 Cost Planning as a Self-fulfilling Prophecy

Raftery (1984, 1987) submits that cost planning is a self-fulfilling prophecy. This statement suggests that once the initial estimate is produced (from which the cost limit or budget is set) that the quantity surveyor will strive to keep the forecasts as close as possible to this figure, thus fulfilling his initial "prophecy" as to what the cost will be. This premise contrasts with the traditional beliefs, which assume that the accuracy improves as the design progresses. In addition, it stresses the critical importance of the accuracy of the initial estimate.

In their study of current literature Ashworth and Skitmore (1983) and Skitmore (1987) conclude that the accuracy of forecasts does not significantly improve between the early and detailed stages of design, thus the value of the forecast does not significantly change. This coincides with Raftery's supposition that the estimated tender figure is effectively set from the first forecast of cost. Since the first forecast is that which usually determines the cost limit or budget, this forecast will be of particular importance to the client who may base all financial decisions on the assumption that the cost of the project will closely resemble this estimated value. In light of such decisions made by the client it is the function of the quantity surveyor to ensure that this cost is not exceeded.

2.6 The Measurement of Accuracy

2.6.1 Introduction

Estimates are produced throughout the course of design, from the inception until immediately prior to tender. However, due to the fact that the design will continuously be revised and changed, the design prepared at the inception will often bear little relation to that immediately prior to tender. It is for this reason that the accuracy cannot be measured over the entire design process, but only from the final estimate. This final estimate is directly comparable with the contractor's tender figure since they are based on the same information and are prepared at roughly the same point in time. However, it is not uncommon for design changes to occur between the time that the final estimate is produced and the tender date which can result in even this estimate being invalid.

2.6.2 The Measurement Datum

When calculating the accuracy achieved, the quantity surveyor's forecast must be compared with a known datum or parameter. This datum may be the lowest tender, the second or third lowest tender, the mean of the tenders, the median of the tenders, a "reasonable" tender figure, or even the final account figure, depending on the author or practitioner's point of view.

Morrison (1983, 1984), Stevens (1983), Raftery (1987), Tan (1988) amongst others all suggest that the lowest accepted tender figure be used to determine accuracy. Since, in most instances the client will accept the lowest tender (except in the case where it is apparent that the lowest bid is an error) it would appear logical that this is the most useful of all bids to attempt to predict. Beeston (1975, p.144) concurs, stating that "the practical use of aiming the estimate at any but the lowest tender is obscure" since the "usefulness of estimating them is doubtful".

Since accuracy and error have been defined, their calculation and measurement can now be explained. The measurement of the accuracy of a price forecast involves mathematically or statistically expressing the degree of deviation of the forecast from the contractor's tender. Since both the quantity surveyor's and the contractor's estimates are variable and subject to error, the variability of one will directly influence the accuracy achieved. Accuracy can be measured in terms of :

1. Percentage error.
2. Mean percentage error.
3. Mean deviation or mean absolute error.
4. Standard deviation.
5. Coefficient of variation of tenders.
6. Coefficient of variation in price forecasts

2.6.3 Percentage Error

This is calculated as follows :

$$\text{ERROR} = \frac{(\text{ESTIMATE} - \text{LOW BID})}{\text{LOW BID}} \times 100\ (\%)$$

This expresses error (or accuracy) as a percentage and is useful when comparing the accuracy achieved in two similar projects.

2.6.4 Percentage Mean Error

This is used to determine the average error of a series of error observations in a sample. It is calculated by adding the percentage errors (as described above) between the sets of variables (i.e. the price forecasts and lowest bids), and dividing by the number of observations in the sample.

$$\text{MEAN \% ERROR} = \frac{\text{SUM OF \% ERRORS}}{\text{NUMBER IN SAMPLE}}$$

Since the signs (i.e. positive or negative) of the percentage errors are taken into account, a 10% overestimation will cancel out a 10% underestimation. This measure of error (or accuracy) is used to detect whether the deviation of one of the variables displays bias or whether their average deviations are equal. Due to the fact that signs are taken into account in the calculation of mean percentage error, the resultant percentage does not establish the size or the variability of the deviations between the variables. To overcome this the mean deviation and coefficient of variation are calculated.

2.6.5 Mean Deviation

This calculation is similar to that for mean percentage error, except that the signs of the percentage errors are ignored, so that a 10% overestimation and 10% underestimation would result in a mean deviation of 10%. This measure can also be termed mean absolute percentage error.

2.6.6 Standard Deviation

The standard deviation is used to measure the dispersion of results around their mean value. It overcomes the limitation of the previous three measures, which are merely averages that do not give any indication of the total spread of the deviations that make up the average value. Standard deviation is measured by calculating the square root of the mean of the sum of the squares of the deviation.

2.6.7 Coefficient of Variation

The coefficient of variation is the most frequently used measure of accuracy by researchers. It is calculated by dividing the standard deviation of a sample by the mean value of that sample and is expressed as a percentage. It is thus a ratio of the standard deviation to the mean. The coefficient determines the consistency of a series of observations, but not, strictly speaking, the accuracy. It can be used to calculate the variability of the tenders or price forecasts using the ratios of the forecasts to the low bids.

2.7 Uncertainty and Risk

2.7.1 Introduction

Uncertainty and risk are intrinsic components of the price forecasting process. Their consequences can never be fully anticipated nor eliminated, only measured and evaluated to determine their probable effect on the accuracy of the price forecast. Thus, all factors that are subject to uncertainty in the forecast will directly affect its accuracy. Whittaker (1970) cited by Skitmore (1981), regards uncertainty to be a consequence of "information gaps and the subjective nature of estimating".

2.7.2 Definitions of Risk and Uncertainty

Bowen and Edwards (1985) define risk as "the extent to which the actual outcome may diverge from what is expected". Risk relates to the occurrence of major low probability events, for example a one in a hundred year flood (Whittaker, 1973) cited by Ogunlana and Thorpe (1991). The occurrence of such events results in substantial cost increases. The assessment of risk is usually based on historical data or experience (Flanagan and Stevens, 1990). Uncertainty refers to the probability of more likely events occurring which can be reasonably anticipated. Flanagan and Stevens (1990), due to the similarities of the terms use them synonymously, and Hertz and Thomas (1983) cited by Birnie and Yates (1991), define risk as meaning uncertainty. For the purposes of this dissertation they are deemed to be the same and are used interchangeably.

Risk and uncertainty, being inherent in every procedure that involves the prediction of a future event, will thus affect the accuracy of both the design price forecasts and the contractor's estimate.

2.7.3 Measurement of Risk and Uncertainty

The supposition that an increase in the level of precision with which quantities and unit rates are calculated will result in a reliable forecast is erroneous, since the prices (both historical in the form of data as well as those being forecast) are intrinsically variable. The quantity surveyor needs to measure the degree of variability and uncertainty of the forecast to determine the probability of that forecast being accurate. Most of the price forecasting techniques do not recognize, nor attempt to measure their variability, which results in them being inherently flawed. Flanagan and Norman (1982a) and Birnie and Yates (1991) describe the use of Monte Carlo Simulation as the most suitable means of identifying :

a) the probability that the contractor's tender price will not exceed the prediction, and
b) the most likely range within which the contractor's tender price will lie.

A full explanation of Monte Carlo Simulation is included in section 2.9.

2.8 The Variability of the Contractor's Bid

2.8.1 Introduction

Due to the fact that the accuracy of the quantity surveyor's forecast is measured relative to the contractor's bid, the accuracy of the accepted tender will determine the accuracy of the forecast. Thus the accuracy of the design forecast is a function of the accuracy of the accepted tender.

Due to the fact that the contractor's estimate is subject to subjectivity, uncertainty and error; the accepted tender figure of a contract will not necessarily (if ever) indicate the true or "actual" cost of the project. This figure will merely represent the contract's present market value. Research carried out by Ashworth, *et al.* (1980) indicates that the contractor's estimated tender figure can be as much as 27% higher than the "actual" cost. The accuracy achieved by the contractor is determined by assessing the precision with which she/he can estimate this "actual" cost.

2.8.2 Determinants of Variability

Skitmore (1981) ascribes the dispersion of tenders to be a result of the variability found in both the components of the tender, i.e. cost estimate and its mark up. The cost estimate's variability is attributed to (Skitmore, 1982) :

1. Inherent unpredictability (e.g. weather conditions, site performance)
2. Uncertainty due to incomplete design and future cost levels
3. Costing errors
4. Subjective allowances for contingencies.

Lipson (1987, cited by Drew and Skitmore, 1992) contends that the following factors will determine the type and degree of mark up (and thus the mark up variability) :

1. Work in hand
2. Bids in hand
3. Availability of staff
4. Profitability
5. Ability of the architect or other supervising officer
6. Contract conditions
7. Site conditions
8. Construction methods and programme
9. Market conditions
10. Identity of other bidders.

Skitmore further encompasses all of the above factors in terms of the bidder's degree of competitiveness (i.e. the desire to obtain the contract). Collusion between contractors and the submission of "cover prices" will also affect the dispersion of the bids. Skitmore (1981) concludes that the prevailing market conditions are the predominating factor which influence the mark-up chosen.

2.8.3 The Variability of Tenders

A large number of authors have, through a series of empirical studies, shown the coefficient of variation of tenders to range between 5 and 7%. These authors include Beeston (1975, p.142), who determined the variability to be 5.2%, which he increased to 6% due to "the recent uncertainty in the market". The following lists the variability measured (in terms of the coefficient of variation) by a number of authors :

* Barnes (1971; cited by Skitmore, 1981) showed a CV of 6.5%,
* Fine and Hackemar (1970; cited by Ogunlana, 1989) 5%,
* Gates (1967; cited by Ashworth and Skitmore, 1983) 7.5%,
* Benjamin and Meador (1979; cited by Runeson, 1988) 6.6%,
* Gryner and Whittaker (1973, cited by Ogunlana, 1989) 6.04%,
* Runeson 4.9%, Keating (1977, cited by Ogunlana and Thorpe, 1987) 5%.

2.8.4 The Effect of the Variability of Tenders

By determining the variability of the contractors' tenders, it is possible to determine the highest potential level of accuracy that the quantity surveyor can achieve. Beeston (1975, p.143) concludes that it is "difficult to achieve further improvement without helping the contractor to reduce his own estimating variability".

2.9 Expectations of Accuracy

Skitmore (1990) contends that the clients' expectations of the degree of accuracy are influenced by their perceptions of usefulness of the forecast. Satisfaction is a function of this perception of usefulness. Thus the satisfaction that the client receives from the quantity surveyor's services will be influenced by their expectations of accuracy of the forecasts prepared. Grieg (1981) cited by Ogunlana (1989), suggested that clients were not strongly dissatisfied with the level of accuracy achieved by quantity surveyors.

In a survey conducted by Bowen (1992), clients and architects were asked what they expected the accuracy levels should be at each of the stages of design. The following table is adapted from the responses of this survey. The percentage ranges given in the table on the following table are those percentages stated to be expected by the majority of those questioned, and are not average values.

TABLE 2.1 CLIENT AND ARCHITECT EXPECTATIONS OF PRICE FORECAST ACCURACY AT EACH STAGE OF THE DESIGN PROCESS

Design Stage	Client	Architect
Inception	5 - 10%	10 - 15%
Feasibility	< 5%	10 - 15%
Sketch Design	< 5%	5 - 10%
Detail Design	< 5%	< 10%

Ogunlana (1989), in an opinion survey of 51 quantity surveying offices established that the accepted level of accuracy should be at least within 15% of the bid at tender stage. Holm (1985, cited by Bowen, 1992) concluded that clients and architects were satisfied with the services provided by the quantity surveyor, suggesting that the quantity surveyor's estimates were adequate in terms of the levels of accuracy expect from them.

2.10 Price Forecasting Techniques

2.10.1 Introduction

Cost models or price forecasting techniques may be defined as "a symbolic representation of a system, expressing the content of that system in terms of the factors which influence its cost" (Ferry and Brandon, 1991, p. 105). In other words these methods represent the cost significant items in such a way as to allow the prediction of the cost to be undertaken, taking into account other non-measurable variables specific to that project such as project type, size, duration, etc. The various cost models are merely different procedures by which an estimate may be prepared.

The performance of a price forecasting model can be assessed in terms of the following criteria :

1. Is it appropriate to what is being measured?

2. Is it appropriate to when it is measured?
3. Is it explicit with regard to measured fact, assumption, opinion and currency?
4. Does it contribute adequately to the cost control process?
5. Is it explicit with regard to uncertainty?

2.10.2 Price Forecasting over the Design Process

The use of a particular cost model or price forecasting technique will be dependent on (Ogunlana and Thorpe, 1987) :

1. the amount of cost data available to the forecaster;
2. the time available to prepare the estimate;
3. the purpose for which the forecast is required; and
4. the amount of design information available.

These factors are, in turn, largely dependent on the degree to which the design has evolved; making the estimating technique chosen by the quantity surveyor theoretically a function of the design stage. Ferry and Brandon (1991) illustrate, in the diagram on the following page, how the different cost models are theoretically used in conjunction with the development of the design.

DESIGN STAGE	TRADITIONAL MODEL TYPE	EXAMPLE
Brief Stage / Sketch Design	Unit	Cost/bed Cost/seat
	Space	Cost/m2 GFA Cost/m3 volume
Detail Design	Element	Cost of element as m2 of GFA
	Feature	Approximate Quantities
Working Drawings	S.S.M. Items	Bills of Quantities

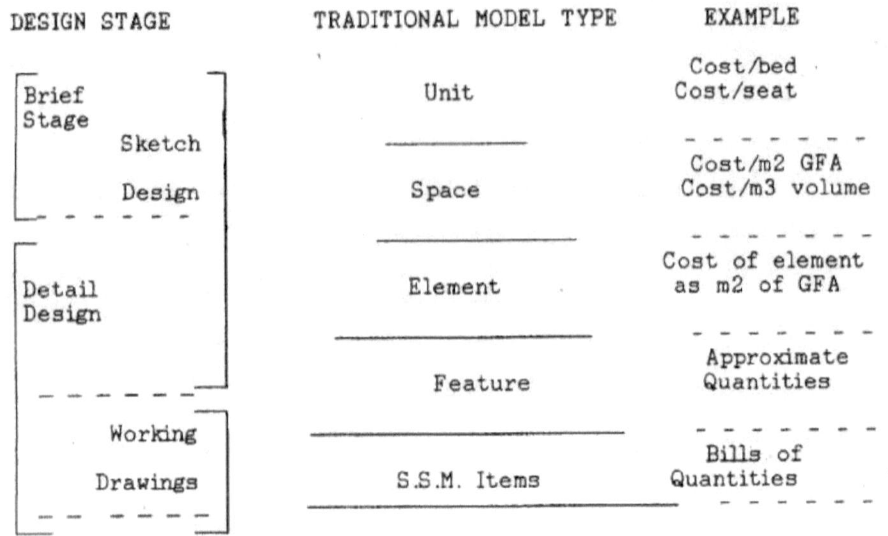

FIGURE 2.3 TRADITIONAL COST MODELLING TECHNIQUES

(Adapted from Ferry and Brandon, 1991)

At the inception stage of the design process there is little, if any, information available relating to the design. As it gradually develops, more details are made available to the quantity surveyor. Since the various price forecasting techniques are of differing degrees of complexity, the level of the detail of information which each requires will differ. It follows that the quantity surveyor should choose a model which requires a level of design detail equal to that currently available.

Raftery (1984) describes this in terms of the design/data/model interface. He contends that the choice of model used should be matched with the degree of development of the design. In addition, the source and nature of historical cost data must be suitable for the model in which it is to be used. All of these factors (i.e. design information available, cost data and forecasting technique used) are recognized as contributing in some way to the accuracy of the estimate. Raftery (1984) represents the design/data/model interface by the following illustration.

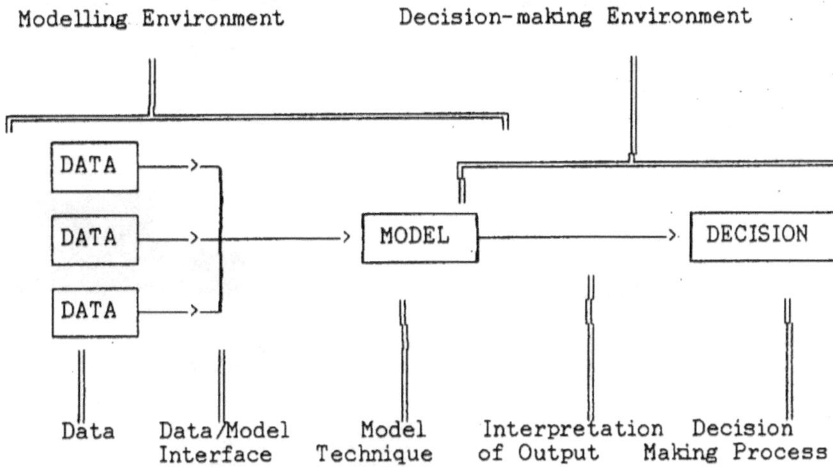

FIGURE 2.4 THE DESIGN/DATA/MODEL INTERFACE

(Adapted from Raftery, 1984)

Figure 2.5 on the following page shows the degree of use of eight estimating methods at each of the design stages. The design stages, labelled 1 to 7 correspond with stages A to G of the Royal Institute of British Architects (RIBA) Plan of Work.

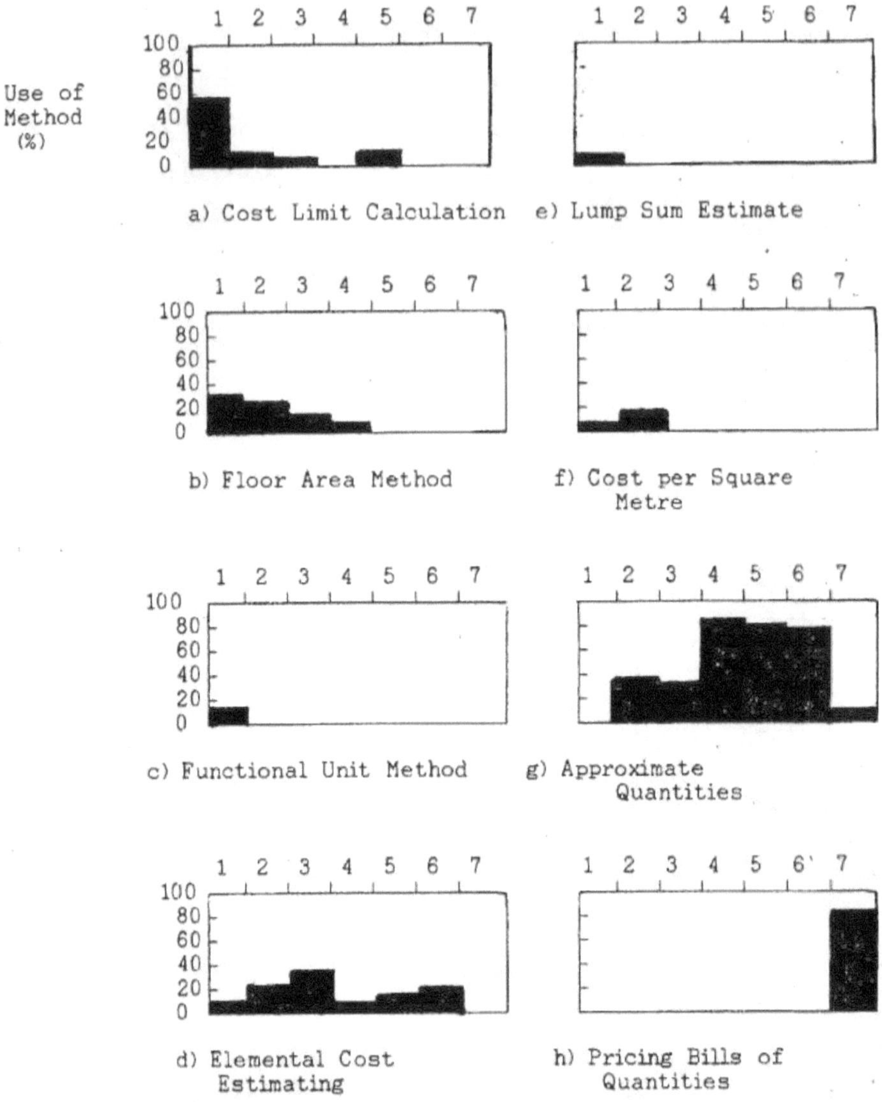

FIGURE 2.5 THE RELATIONSHIP BETWEEN THE DESIGN STAGE AND THE ESTIMATING METHOD USED

(Property Services Agency, 1980)

Property Services Agency (1980) compiled the above figure from a survey of six UK quantity surveying offices. It can be clearly seen that at all stages of the design except Inception (stage 1) and Bills of Quantities Stage (stage 7), that the use of the approximate quantities method predominates. It was concluded that quantity surveyors use the most detailed method of estimating available to them based on the data at hand.

2.10.3 Traditional Techniques

The "traditional" price forecasting techniques are categorised into single and multi price-rate methods.

2.10.3.1 Single Price-Rate Methods

The single price-rate methods include :

a) The Unit Method

This method is based on the assumption that there is a close relationship between the total cost of the structure and the number of functional units it accommodates. The cost is often expressed in terms of the cost per functional unit, this is then multiplied by the number of these units to determine the total cost. The type of functional unit used, logically, will depend on the function of the building. For example the cost of a school would be expressed in terms of the cost per pupil, while the cost of a hospital in terms of the cost per bed. Since the unit costs of previous projects are used as a basis for the estimated cost, there must be a standard quality of construction and finish for this technique to be effective. This method is useful since it is relatively simple and can be used when little or no design information is available. However, unit rates are prone to become obsolete over a short period of time. In addition they are also difficult to calculate as the differing design variables (such as size, shape, etc.) must be incorporated into this rate. Even a small error in the unit rate is magnified significantly, resulting in this method being potentially extremely inaccurate.

b) The Cube Method

This method was only used in the UK, and is now largely obsolete. A standard code of practice was developed which embodied a series of rules for measurement. The volume of the structure was measured which was then multiplied by a rate per cubic metre. The historical rates used were frequently inaccurate due to the inherent dissimilarities of structures. Slight errors in the rates have an even larger impact than those for unit rates, since the rate is multiplied by a cubic area rather than a square area.

c) The Superficial Area Method

This method is commonly used. The price of the building is expressed in terms of the cost per square metre. The area of the building is measured to the outer faces of the walls, (in South Africa only - the area is generally measured to the inner faces of wall in the U.K.) keeping different forms of construction separate. Other items which are not related to the size of the building are added separately. Standard rates can then be determined for each type of building, e.g. offices, housing, etc. However it is difficult to accurately take into account the effect of storey height, plan shape, density, quality of finishes, etc. in the square metre rate.

d) The Story Enclosure Method

The South African storey enclosure method differs somewhat from that practised in the UK (as described by Ashworth, 1988). It was developed by the Public Works Department (PWD) due to the fact that much of the work carried out by this department is similar in nature. It is a consolidation of the approximate quantities and superficial area methods. The measurement is compared to that of the "quasi-house" as defined by the PWD. The seven prescribed component parts are measured and then multiplied by the appropriate tariff and factor. The tariff is an index which sets the pricing level. The factors are constant and are predetermined for each item. It is claimed (Brooker, 1991) that a high degree of accuracy can be achieved, but due to little understanding and misuse this level is not always attained.

2.10.3.2 Multi Price-Rate Methods

The multi price-rate methods are significantly more detailed in their approach. Such methods include :

a) Approximate Quantities

The approximate quantities method involves the measurement of quantities in much the same way as a bill of quantities. However, as the name implies, the quantities are not measured precisely and items of equal measurement are frequently combined into single items. The time taken to prepare this type of estimate is substantially longer than the previous methods due to the degree of detail with which the items are measured. Due to the high degree of detail, specifications often have to be assumed which may be inaccurate. Property Services Agency (1980) found that in the UK, approximate quantities were the most commonly used method of estimating at all stages of the design except inception and tender.

b) Elemental Estimating

In South Africa, elemental estimating is based on the "Guide to Elemental Cost Analysis". This document defines the scope of the measurement of the functional elements and components. Although this guide was devised for cost analyses, it is used for estimates which are based on data derived from such analyses and are prepared in much the same format. The preparation of an elemental estimate involves the updating of the analyses for time (using the BER indices) and the adjustment for quantity and quality differences. This method has the advantage of showing the distribution of the costs amongst the elements which helps to ensure an even allocation of funds.

c) Pricing the Bills of Quantities

As the final check to ensure that the costs are within the budget, the Bills of Quantities are frequently priced immediately prior to tender. The Bills of Quantities are prepared in accordance with the provisions set out in the Standard System of Measurement published by the Association of South African Quantity Surveyors.

2.10.3.3 Criticisms of the Traditional Forecasting Techniques

The so-called traditional forecasting techniques have been widely criticized by various authors (Bowen and Edwards, 1985; Mathur, 1982; Bowen, *et al.*, 1987; Beeston, 1987) amongst others. The flaws of these techniques have largely been attributed to (Bowen and Edwards, 1985) :

1. Determinism : The traditional approaches rely on the use of cost data which is derived from historical sources. The nature of this data has proven to be inherently variable (Ashworth *et al.*, 1980; Beeston, 1975) and uncertain. The use of such data without statistical qualification of its inherent variability makes these techniques deterministic in nature.

2. Inexplicability : The traditional cost models make no attempt to provide any explanation of the system which they claim to represent, since the estimated total cost is based on

the finished product rather than the actual process. Bowen and Edwards (1985) contend that the models should be "logically transparent" in the manner in which they account for the cost implications of the construction process, and thus should seek to simulate it.

3. Unrelatedness : The interdependency and relationships between the costs of the elements and components are not fully understood and have not been considered by the traditional techniques. As a result, Raftery (1987) contends that linear or "straight line" relationships are incorrectly assumed to exist between the cost and quantity or quality of the variable.

2.10.4 Non-Traditional Techniques

Due to the above shortcomings, Bowen and Edwards (1985) suggest a "paradigm shift" from the traditional cost models to forecasting techniques which will replace historical determinism with stochastic variability, inexplicability with logical transparency and unrelatedness with interdependence. The use of so-called "non-traditional" forecasting techniques have been an attempt to do so. The majority of these techniques, although they can be used for estimating purposes, are not strictly price forecasting techniques themselves. However, since they can be used for forecasting they are discussed below.

2.10.4.1 Regression Analysis

Regression analysis seeks to establish relationships between variables and to determine formulae which will best describe these relationships. Since exact relationships between variables seldom exist, average relationships are determined statistically by plotting the data relating to the relationship, and using the method of "least squares" (Ferry and Brandon, 1991) to find the line of best fit. The equation of this line forms a mathematical representation of the relationship between the variables. However, this technique does not overcome the limitations of determinism (since it does not quantify uncertainty), and unrelatedness (as it does not consider the interdependency of elements) associated with the traditional techniques. In addition, (Skitmore and Patchell, 1990) argue that an extremely large data base is required to ensure "reasonably robust" results.

2.10.4.2 Monte Carlo Simulation

Monte Carlo simulation is a means of sampling from a distribution in a random manner to provide a range of solutions from which the optimal solution may be established. A distribution is formed from historical cost data for each elemental category and is represented in the form of a histogram which is translated into a cumulative distribution curve. The sampling of a random value from each elemental distribution takes place during each simulation. The sum of all the values from a single simulation multiplied by their respective quantities gives the total cost. A probability distribution is formed from the total costs generated by the simulations. From this distribution the total project cost is sampled. This forecasting technique takes uncertainty into account, overcoming determinism. However, its major flaw (Wilson, 1982) is the fact that it assumes that all elements are independent (i.e., their costs are unrelated to one another) and that the probability distributions are normally distributed. Wilson (1982) however, suggests that they are skewed to the right.

2.10.4.3 Geometric Optimization

Geometric optimization is merely a refinement of the traditional techniques. It seeks to express the building in algebraic terms, then to give values to those terms and to apply a cost co-efficient, which determines the total project cost. The algebraic equations are based upon observation, experience and intuition. One of the main advantages of this technique is its flexibility. However, it does not overcome any of the problems associated with the traditional models. Difficulties are also experienced in determining algebraic formulae to represent the building, and choosing the appropriate cost factor to be applied.

2.10.4.4 Cost Mapping

Brandon (1984) was the first to introduce the concept of cost mapping. Bowen and Wolvaardt (1988) describe cost mapping as the technique of superimposing cost "contours" (i.e. the lines of equal cost) on the cost/parameter tables to reflect the sensitivity of cost to changes in the parameters. These parameters are design variables (such as plan shape, storey height, etc.). An optimization technique (such as linear programming) is then used to determine the optimal value of the design parameter which will establish the optimal cost. This method has had relatively limited acceptance by practitioners.

2.10.4.5 Expert Systems

An expert or knowledge based system is defined by Newton (1984, p. 11) as "a computer system which embodies the essential knowledge for any particular function (for example estimating) ... in such a way that reflects the decision-making process of human specialists in that field". It consists of a "knowledge base", which incorporates the expert's knowledge usually in the form of rules, and an "inference engine" which mimics the thought processes of the human mind to determine logical solutions. Characteristically an expert system is able to reason with judgemental knowledge, explain its line of thinking (i.e. is explicable), can cope with uncertain information in a probabilistic rather than deterministic manner, can expand its knowledge base and deliver it output as advice instead of just facts. It thus overcomes the limitations of determinism and inexplicability. The Royal Institute of Chartered Surveyors (RICS) helped to develop the first commercially available expert system package (named ELSIE) which was tailored especially for the use in the construction industry.

2.10.4.6 Resource Based Cost Models

Property Services Agency (1980); Beeston (1987); Bowen, *et al* (1987); Ferry and Brandon (1991), amongst others contend that one of the major weaknesses of the foregoing forecasting models is that they are based on "in-place quantities" (i.e. the finished product) and not the actual construction process. This results in a mismatch of the process biased data with a design biased model (Bowen and Edwards, 1985). It is claimed (Beeston, 1987) that techniques based on "in-place quantities" have reached the limit of their development and that the accuracy achievable is inadequate for modern business practices. Raftery (1984) criticizes "in-place" quantities methods for their reliance on historical cost data. He contends that since each building is unique and built only once, it cannot rely on the comparisons with other buildings (i.e. for the use of historical cost data), making such methods inherently flawed. All of the foregoing models rely to some extent on such data. The above mentioned authors advocate the use of contractors estimating methods for price forecasting purposes. Bennett (1978) and Ogunlana (1989) identify the estimating methods used by contractors to calculate their rates as unit rate estimating, spot rates and operational estimating.

2.11 Historical Cost Data Used

All of the price forecasting techniques discussed in this chapter require some degree of historical cost data upon which the estimate will be based. The quality of this data is of critical importance since the accuracy of the forecast is entirely dependent on the upon the costs applied to it (Ferry and Brandon, 1991).

2.11.1 The Nature of Cost Data

Ashworth (1988) describes cost data as being hierarchical in nature, structured according to its level of detail. Eight levels are identified, the total contract sum being the level of least detail and basic labour rates the level of greatest detail. The nature of the cost data required depends on the price forecasting technique used.

2.11.2 Sources of Cost Data

The sources from which cost data may be obtained can be categorised into two types :

1. "in-house" data
2. published information

"In-house" cost data is that which is produced within the quantity surveyor's own office and stored in the form of priced Bills of Quantities and cost analyses. Property Services Agency's (1980) study in the UK showed that quantity surveyors have a clear preference for using "in-house" data, particularly those projects with which they were personally involved. Only when this source was inadequate did they consult published media. The reason for this is that they are familiar with and have a full understanding of the data with which they are working. They are thus able to relate it to new projects (with or without any adjustments) with greater confidence. Morrison (1983), in his analysis of UK quantity surveying firms found that the majority of those questioned did not maintain a formal cost library of "in-house" data. He concluded that they prefer to use a single source of cost data which is usually a priced bill of quantities instead of an average rate obtained from a library.

In the UK there are numerous sources from which published cost data can be obtained. The Building Cost Information Service (BCIS) operates a computer-based information service which supplies elemental cost analyses and cost studies to each member. Various price books are also available including Spon's Architects and Builders Price Book, Laxtons Price Book, Hutchins Priced Schedules and Griffith's Building Price Book, amongst other.
In South Africa the choices are far more limited. No

information service exists, and the only comparable price book is "Merkels Builder's Pricing and Management Manual". The Department of Statistics produces a document which contains average rates for many of the items measured in the Bills of Quantities according to the region, building type and quarterly period. However this document is only published at infrequent intervals and is often more than six months out of date by the time it is published. The Bureau for Economic Research (BER) also publishes a handbook which provides pre-contract escalation indices as well as average prices for twenty-two building components. Other building price indices include those published by Department of Statistics (civil engineering indices) and the BIAC "Haylett" indices, as well as a forecasting service provided by Medium Term Forecasting Associates (Pty) Ltd.

2.11.3 Suitability of Cost Data

Raftery's (1984) design/data/model interface explains the need for the cost data to match the forecasting model used (in terms of its level of detail), which in turn should correspond with the appropriate stage of design. Cost data, design information and forecasting models are all hierarchical in nature, progressing from less refined to highly detailed as the design develops.

2.11.4 Reliability of Cost Data

The cost data from a particular building are a reflection of the circumstances peculiar to that project and when such data is applied to any other project will make it inherently flawed.

Raftery (1984) describes the two major transformations of this data. The first occurs when the contractor prices the Bills of Quantities. The rates which he inserts are his subjective assessments of the costs which are subject to tactical decisions, such as "forward loading". These rates will not represent the actual costs of the items and are merely notional breakdowns of the overall price. Beeston's (1975) analysis of the "variation of whole building prices" (p.140) suggested that the variation in the prices of "identical buildings" is 8.5%. He also determined the coefficients of variation of selected rates from bills of quantities for similarly described items for the different trades. The "Excavator" trade showed a variation of up to 45%. From Beeston's findings it can be seen that the rates contained in the bills of quantities are highly variable and uncertain, making them largely unreliable for use in cost estimates. Ashworth, *et al* (1980) confirmed such variances in their study which revealed that the highest estimate (of a brickwork section of nine projects) was 62% above the lowest. These two studies indicate the wide variability of the rates found in priced bills of quantities.

The second transformation of the cost data occurs when the rates are gathered and allocated over the building elements and components in the production of an elemental cost analysis is produced. Bennett, *et al* (1981, p. 35) conclude that "present methods for producing and using cost data taken from bills of quantities are neither sufficiently flexible in use, nor capable of supporting accurate cost estimating."

CHAPTER THREE : THE FACTORS INFLUENCING ESTIMATING ACCURACY

CHAPTER THREE

THE FACTORS INFLUENCING ESTIMATING ACCURACY

3.1 Introduction

The price forecasting process is, by definition, inherently uncertain. Logically, each aspect of the forecast that is subject to uncertainty influences the accuracy of that forecast. It is the level of this uncertainty which determines the degree of accuracy achievable.

Ogunlana (1989) represents the variable aspects of a construction project's environment by means of the illustration on the following page. He maintains that uncertainty is derived from the variables emanating from the construction task and its "immediate" and "external" environment. He considers such variables to include the type, size, location and duration of the project; the state of the economy, resource availability, etc. It is the function of the estimator to make realistic allowances in the forecast for these factors in anticipation of their effect on the contractor's bid. A discussion of the effects of these, and other factors, on forecasting performance follows.

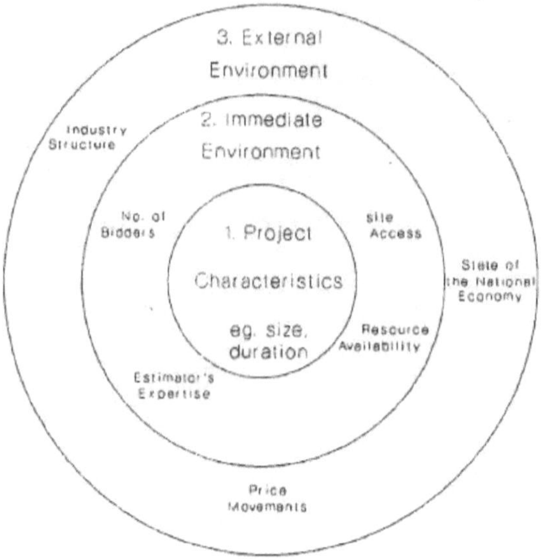

FIGURE 3.1 THE UNCERTAIN NATURE OF CONSTRUCTION PROJECTS

(Adapted from Ogunlana, 1989)

3.2 Type of Project

Projects may be categorised into various types according to the function of the structure; for example schools, offices, housing, factories, health centres, etc. It is thought (Ogunlana, 1989) that the repetitive nature of some types of projects makes the estimating task more analytical in nature, and thus their costs will be easier to assess. This assumption appeared to be confirmed by McCaffer's (1976) analysis of 132 Belgian public building projects and 168 public roads projects. The building contracts (5.2% underestimation) were on average less accurately estimated than the roads projects (1.5% underestimation). Although, the roads projects exhibited a higher standard deviation indicating less consistency in the forecasts.

Harvey (1979) analysed 2401 Canadian public works contracts, which she categorised into building (944 projects), non-building (i.e. roads, bridges, etc.; 633 projects), special trades (i.e. electrical, plumbing, etc.; 743 projects) and "other" projects (203 projects). The results from the analysis of the standard deviations and CV's of each type showed that the estimates were generally slightly higher for non-building and lower for special trades, as opposed to building projects.

Property Services Agency (1980) studied data from 958 projects obtained from six UK public sector quantity surveying offices. It was concluded from this data that accuracy tended to improve on housing and school projects. In contrast, Flanagan and Norman's (1983) analysis of 166 County Council projects, found that school projects did not exhibit a significantly improved level of accuracy. Skitmore (1988) examined 33 local authority building projects confirming Flanagan and Norman's deduction. Although the results indicated that schools and offices were more accurately estimated, the analysis of variance could find no statistically significant relationship. Skitmore's (1987) experiments with twelve quantity surveyors found the differences between the mean errors of different types of projects to be statistically significant (Health Care 15.14%,

Offices 24.76%, Schools -11.11%, Housing -7.52% and Factories 19.89% mean error). Betts and Gunner's (1989) analysis of 46 Singapore construction contracts also suggested that certain types of projects exhibited a higher level of accuracy than others. The improved accuracy level of was attributed to the experience or expertise of the surveyor with these types of projects.

Bennett, *et al.* (1981), in a study of 915 projects from seven offices (related to the research carried out in 1980), found accuracy to be significantly higher on supermarket developments schools and housing projects for some of the offices. They attributed the higher level of accuracy to the fact that these types of projects constituted the greater proportion of that office's work load, suggesting that the improved accuracy was as a result of experience with this type of project. Both Bennett, *et al's* and Betts and Gunner's research suggests that the apparent relationship between building type and accuracy is as a result of experience, implying that the type of project would have little influence on accuracy where the forecaster has limited or no experience.

3.3 Duration of the Project

Ogunlana (1989) contends that as the duration of a project increases, so will the likelihood of changes occurring. He suggests that the forecast's accuracy will be negatively affected since such variations are difficult to accurately anticipate or adjust for. It is implied that as the contract duration increases, it is likely that the accuracy will decrease. Ogunlana (1989) identifies two sources of variations in a project :

1. External changes in the project environment due to changes in the state of the economy which will affect the prices and availability of labour and materials.

2. Design changes as a result of changes in taste. Merrow, *et al.* (1979) cited by Ogunlana and Thorpe (1991) observed that changes in the scope of a project were more frequent on contracts of longer duration resulting in a significant underestimation of costs.

Skitmore's (1988) analysis of 33 UK government and 67 American government projects indicated a general trend of increasing accuracy with longer contract durations. However these trends were not shown to be statistically significant. In contrast to this, Tan (1988) concluded, from an analysis of the same 33 UK projects that, although projects of longer duration were more consistently estimated, that the accuracy for longer projects deteriorated as the contract period increased. Tan's calculations of mean absolute accuracy did not appear to co-incide with her deduction. The analysis of the 67 USA projects by Tan confirmed Skitmore's observations.

3.4 Size or Value of the Project

The accuracy achieved may be affected by the size or contract value of a project in the following ways :

1. Since the degree of error contained in the forecast is expressed as a percentage of the contract sum, then the higher the contract sum the lower the percentage error of a fixed rand amount will appear. For example an error of two out of five is 40%, while two out of ten is only 20%.

2. It is submitted (Ogunlana, 1989) that the Central Limit Theorem, which results in the cancelling out of an overestimation by an underestimation, is more likely to affect a larger project than a smaller one. Thus the probability of errors being cancelled out is thought to be greater on larger projects.

3. Ogunlana and Thorpe (1991) contend that only larger projects, due to the costs involved in estimating, can justify the time devoted to preparing detailed (and thus more accurate?) estimates.

4. In contrast to the above views, which imply that higher contract values may result in improved accuracy, it is contended (Ogunlana,1989) that larger projects frequently have a higher degree of uncertainty.

5. Based on Runeson's (1988) analysis of tender distributions, which shows that the average number of bidders for a

contract increases as the value increases, Ogunlana (1989) suggests that a higher contract sum results in a greater variability of the tenders. Beeston (1975), Stevens (1983), and Morrison (1984), amongst others, have concluded that the extent of the variability of tenders will directly determine the accuracy achievable. Since the quantity surveyor's level of cannot theoretically be lower than that of the tenderers. However, Ogunlana's (1989) interpretation of Runeson's research suggesting a larger number of bidders results in a higher variability of their tenders, is totally unsubstantiated. It contradicts the findings of McCaffer (1976) who showed variability of tenders to decrease as the number of tenderers increased.

McCaffer (1976), from the analysis of 300 Belgian building and roads contracts, concluded that there was no correlation between contract size and accuracy. Wilson *et al.* (1987) in their analysis of 408 Australian government contracts, indicated that the larger the value of the project the lower the accuracy achieved. The results however, did not show any statistically significant correlation between size and accuracy.

Harvey (1979), by means of regression models, showed accuracy to improve with increasing contract value. Property Services Agency's (1980) analysis of 958 projects confirmed Harvey's assessment. An analysis of 46 Singapore building contracts conducted by Betts and Gunner (1989) also indicated that estimating accuracy improves with increasing project size. This conclusion, however, was drawn from a visual inspection only and no statistical analysis was performed to confirm their assumption.

Flanagan and Norman (1983), by means of a linear regression analysis on 166 projects from two County Councils, found size to significantly influence the estimating accuracy. Council A's data showed tenders were generally overestimated, the percentage overestimation increasing with increased size. Accuracy decreased from 7 to 12%. Council B's data indicated that projects of low value were overestimated (by 2% on average), while those with higher values were underestimated. Accuracy tended to decline with projects of higher value, although the extent and direction of the decline was not universal.

Skitmore (1988) examined 100 British and American government building and engineering projects, the results of which indicated that accuracy improved with increasing project value. An analysis of variation, however, failed to confirm these assumptions. Ogunlana (1989), when testing 51 UK construction projects, could also find no statistical relationship between accuracy and project size.

3.5 Geographical Location

Differences in the prices and availability of labour and materials, trade unions restrictions, and subcontracting services all contribute to the variability of price levels in different regions. Wallace (1977) cited by Ogunlana (1989), observed that due to local labour regulations and building code requirements, building in certain areas is more problematic, which adversely affects tendering levels.

Harvey's (1979) study of 2401 public works contracts across six different regions in Canada showed that the differences in the errors of each region were statistically significant. Skitmore (1990) suggests that the results found in Property Services Agency (1980) and Flanagan and Norman's (1983) research may, although it is not actually stated by these authors, also be attributed to the effect of the locations of the different offices and County Councils respectively studied. Wilson *et al.* (1987) analysed 410 Australian PWD contracts located over four administrative regions. From the results they concluded that the geographical location of a project does not affect the estimating accuracy. Ogunlana's (1989) study of 51 UK projects was also unable to determine a statistical relationship, thus confirming Wilson's deduction.

Gunner and Betts' (1990) analysis of projects from nine locations distributed around the Asian Pacific Rim showed there to be significant differences in the estimating performance for the locations. The authors acknowledge that the differences in the levels of accuracy may partially have arisen due to the contrasting market conditions at each of the locations, the level of imported components, the number of tenderers, staff specialisation and forecasting methods, for which no allowance was made.

3.6 The State of the Market

The prevailing state of the economy can be represented by its position in the business cycle. This cycle maps the changes in demand for goods and services over time. The "peaks" and "troughs" of the cycle indicate the turning points of the phases of expansion and contraction in the economy. The state of the construction industry emulates that of the business cycle.

Thus, when the economy is in a state of recession, the construction industry will follow suit. A recession affects the construction industry by causing a decrease in the demand for construction, resulting in intense competition for the few contracts available. Contractors are more willing to settle for low mark ups or profit margins. Due to the decrease in work, the labour requirements diminish, resulting in increased retrenchments. An increase in the number of insolvencies and buy outs of companies also occurs. A period of recovery in the economy is characterised by a renewed demand for construction. The increase in work available reduces the level of competition, and profit margins subsequently rise. Resources, due to the improved demand, become relatively scarce causing price rises.

The above explains the implications of the state of the economy to the contractor, the effects of which will be reflected in the tender. The quantity surveyor's forecast, which aims to predict the lowest tender, must then anticipate the magnitude of the allowance made by the contractor for the prevailing market conditions. The ability of the quantity surveyor to take into account these conditions will be apparent by the level of accuracy achieved.

De Neufville, *et al.* (1977) cited by Ashworth and Skitmore (1983), noted that there were distinct differences in the accuracy achieved for the projects which occurred during "good" and "bad" years. "Good" and "bad" years being defined as those of greatest and least construction activity. It was shown that the accuracy of the forecasts made in "good" years was, on average, lower than that made in "bad" years. The reason for the differences between the forecasts was attributed to the fact that quantity surveyors were not immediately aware of the changes in the amount of construction activity and its effect on prices. The diagram below illustrates De Neufville *et al's* findings.

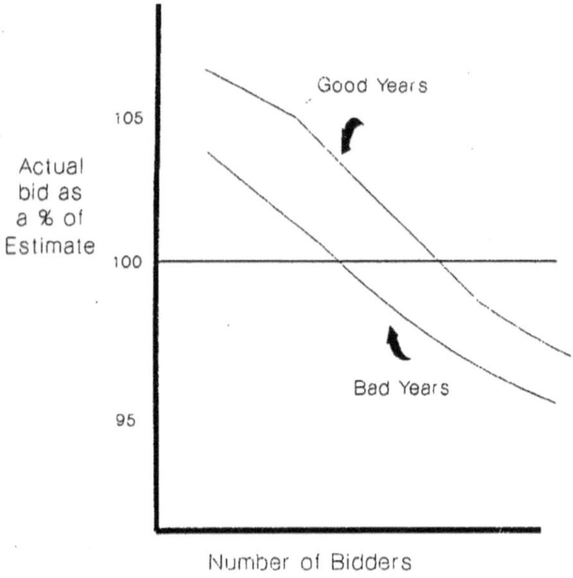

FIGURE 3.2 THE EFFECT OF MARKET CONDITIONS ON ACCURACY

(Adated from De Neufville *et al* 1977, cited by Ashworth and Skitmore, 1983)

Harvey's (1979) analysis of Canadian public works projects confirmed De Neufville *et al*'s theory, showing the accuracy of forecasts to be lower during expansionary phases. The relationship between economic conditions and accuracy was found to vary between the different regional areas, as a result of the different local economic climates.

Property Services Agency (1980), in their study of 958 public sector projects, analysed each office's data according to the year when that the projects were forecasted. Only office A indicated any evidence of market conditions having any effect on its accuracy. The mean absolute errors were found to be significantly larger for the years where they claimed that "there was considerable uncertainty in the building industry" (p 9).

Skitmore (1987b) analysed the effect of the market on price levels. He concluded that the changes in the economic climate, in terms of changes in the price levels of different sizes and types of projects in different geographical locations, was the actual cause of changing accuracy levels.

Betts and Gunner's (1989) study of Singapore construction contracts showed that during times of decreased construction activity, forecasts were on average, overestimated; and during an upswing in the amount of construction activity forecasts were generally underestimated. They attribute this to the fact that the cost data used is historical and thus cannot reflect the changing trends of the market. These observations, gathered from a relatively small amount of data, were not confirmed statistically.

Ogunlana (1989) suggests that the psychological effect of personality (in terms of whether the person is considered an "optimist" or "pessimist") may effect the surveyor's ability to predict market conditions. Hogarth (1981) cited by Ogunlana and Thorpe (1991) observed that growth is usually underestimated, irrespective of the degree of experience or mathematical training of the forecaster. Wagenaar and Sagaria (1975), cited by Ogunlana (1989) found similar results.

3.7 The Number of Bidders

The number of bidders for a contract is determined largely by the degree of competition between contractors in the construction industry as a whole. The level of competition varies depending on the prevailing market conditions, as described in the previous section. Thus the effect of the number of bidders is linked to the state of the market.

Ogunlana (1989) contends that a greater number of bidders for a contract is expected to result in a wider variability of tenders. Consequently it is suggested (Stevens, 1983; Morrison, 1983; Ogunlana, 1989; Beeston, 1975) that a greater variability of tenders may result in a lower level of accuracy since the quantity surveyor cannot estimate with a greater degree of accuracy than the tenderers. However, a small number of bidders may encourage collusion between these contractors, resulting in unexpectedly high tenders and thus an underestimation by the forecaster.
McCaffer (1976), in his analysis of 300 Belgian public works

contracts, established that a negative correlation exists between the low bid/design forecasts ratio's and the number of bids received. Thus a positive correlation exists between the level of accuracy and the number of bidders, i.e. accuracy improves with an increasing number of bidders, which contradicts Ogunlana's (1989) contention. De Neufville *et al.* (1977) studied 167 projects from 1961 to 1974. Their study confirmed McCaffer's conclusion, showing a curved negative relationship between the low bid/design forecast ratio and the number of bidders to exist.

Flanagan and Norman (1983), in an attempt to quantify the effect of market conditions on estimating accuracy, used the number of bidders as an indication, based on the assumption that the number of bidders increased when the market was depressed. Their findings, which showed the degree of error to decrease as the number of bidders increased, substantiated those of McCaffer and De Neufville *et al.*

Wilson *et al's* (1987) analysis by means of a logistic regression also showed the accuracy to improve with a larger number of tenders. The effect of the number of tenders was found to be "reasonably consistent" (Wilson *et al*, p. 222) over all three categories of projects. The results of the studies conducted by Runeson and Bennett (1983) cited by Skitmore (1988), and Hanscomb Associates (1984) cited by Skitmore (1990) confirmed all of the above authors' conclusion.

In contrast to all of the above findings, Ogunlana's (1989) analysis of 51 projects did not show any statistically significant relationship between estimating accuracy and the number of bidders on a project. Skitmore's (1988) study, although detecting a trend of overestimating with an increasing number of bidders, could not show this trend to be statistically significant, thus confirming Ogunlana's conclusion.

3.8 Complexity of the Project

The degree of complexity of the design is determined by the level of unusual, intricate or technical components. It is suggested by Bodily and Hogarth (1981) cited by Ogunlana and Thorpe (1991) that technical complexity reduces the ability to forecast costs accurately. Skitmore's (1988) analysis of 67 building and engineering contracts, indicated that consistency and accuracy declined with increasing complexity. However, these results were not shown to be statistically significant.

3.9 The Level of Design Information Available

As the design evolves, the architect is able to furnish the quantity surveyor with more detailed information, upon which the estimates will be based. The quantity surveyor will, upon the receipt of such information perform subsequent forecasts to ensure that the price remains within the budget. The complexity and degree of detail of the forecasting technique used will increase as more design information is made available (Ferry and Brandon, 1991).

The assumption that the level of accuracy improves once more design details become available has been accepted by both academics and professional quantity surveyors (Skitmore, 1987; Ogunlana and Thorpe, 1991). Barnes (1974) illustrates this assumption by the following figure :

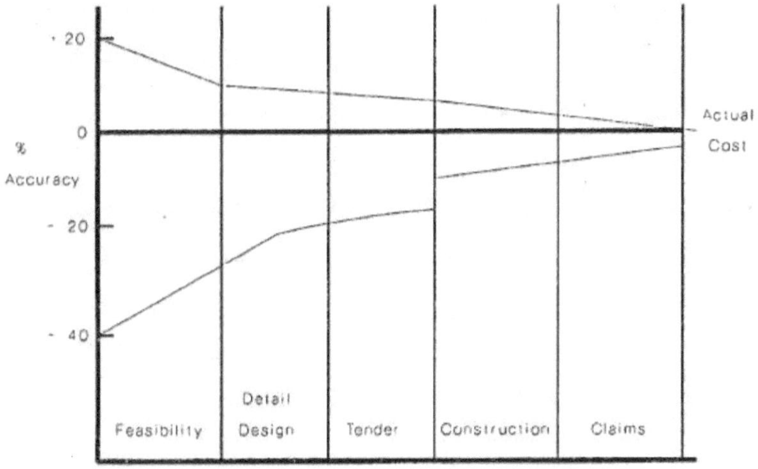

Figure 3.3 PERCENTAGE FORECASTING ERROR OVER TIME
(Adapted from Barnes, (1974) cited by Ogunlana (1989))

Figure 3.3 shows the percentage error to decrease substantially from +20 to -40% at inception to +10 to -20% during the feasibility stage. From the end of feasibility, until the contract is tendered, this level of accuracy improves only slightly. Accuracy is thought to increase over the design process as a result of:

1. better definition and scope of the project being achieved (i.e. there is less uncertainty relating to the construction details) through the acquisition of more detailed design information;

2. once more information is made available, a more detailed estimating techniques can be used, which uses more detailed historical cost data which, is expected to statistically reduce the error inherent in this data (Beeston, 1975).

Ashworth and Skitmore (1983) in a literature review, compared the degree of accuracy measured during the early stages of the design with that at detailed design or immediately prior to tender, to determine the effect that level of design information on accuracy. From the published literature they concluded that the coefficient of variation (CV) improves only slightly from 15 to 20% during the early stages to 13 to 18% during detailed design. This study refuted the theory that accuracy significantly improves over the design process.

The only forecast which is directly comparable with the contractor's tender is that which is submitted immediately prior to tender. This is due to the fact that the design usually undergoes significant changes while it is developed and thus the earlier estimates will not be based on the same design as the contractor's tender. For this reason it is practically impossible to measure the accuracy of any earlier forecasts. Consequently there is relatively little research available which attempts to determine the accuracy of early forecasts, and, much of that which has is of questionable validity. Many of the authors' research pertaining to the accuracy of early stage design price forecasts is based on opinions and not actual measurement. Such authors include Weaver, *et al.* (1963); Park (1972), Barnes (1974), Keating (1977), Marr (1977) and Greig (1981).

To overcome this obstacle Skitmore (1987) conducted experiments with "dummy" projects and twelve quantity surveyors. Five recent building projects were selected at random from an existing data base, from which each quantity surveyor chose two. The design information relating to the projects was divided into sixteen "information types". Each piece of information was provided individually, upon receipt of which the estimate was revised. The results showed that the forecasts of the self-professed "experts" did not improve significantly with the increase of design information available to them.

When provided with only the knowledge of the project type and size, they could estimate the cost with accuracy comparable to that of the "average practitioner" when pricing full bills of quantities. Skitmore (p334) concluded that generally "forecasting accuracy does not improve with additional project information in the manner expected."

3.10 The Price Forecasting Technique Used

In chapter two the various forecasting techniques were discussed. Due to the differences in their approach to estimating, each of these techniques will have a different inherent error (Morrison, (1983)). Morrison (1984) further contends that the degree of error (and thus the level of accuracy) inherent in each technique will be dependent on :

1. the level of detail with which the estimate is performed (i.e. the number of sub-estimates produced), and

2. the variability and nature of cost data used. (The effect of cost data on accuracy is discussed in section 3.11.)

Stevens (1983) supports Morrison's theory that the accuracy achieved is a function of the number of sub-estimates performed. Logically, the variability (i.e. accuracy) of each of these sub-estimates will determine the variability of the estimate. The Central Limit Theorem shows the accuracy of the total estimate to be greater than the average level of accuracy of the sub-estimates (Bennett and Barnes, 1979). As the number of sub-estimates in the forecast increases, the probability of the errors (i.e. overestimation and underestimation) cancelling each other out increases (to a point) which causes the apparently higher level of accuracy. The following graph illustrates the effect of the number of sub-estimates (i.e. the degree of detail of the forecast) on the accuracy of the forecast.

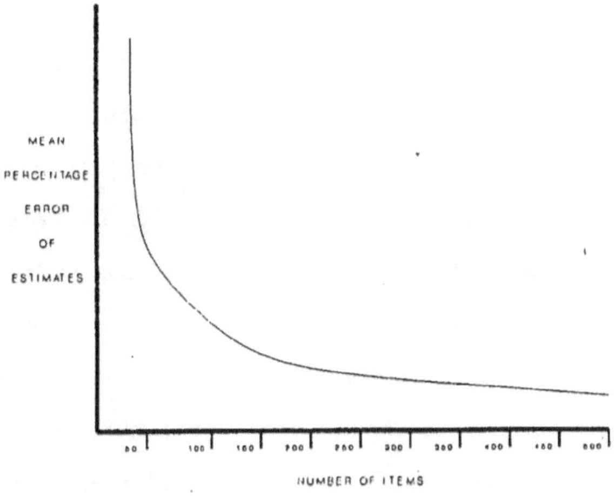

FIGURE 3.4 THE RELATIONSHIP BETWEEN THE DEGREE OF DETAIL OF THE ESTIMATE AND THE ACCURACY ACHIEVED

(Adapted from Stevens, 1983)

From the above figure it can be seen that the percentage error asymptotically approaches zero as the number of items in the estimate increases. Therefore the ratio of the estimate's accuracy to the number of items will continuously decrease, so that an increase in detail results in a declining increase in accuracy. Property Services Agency (1980) sought to determine the optimal number of items which should be measured. This would be the point at which any further sub-estimates would not result in a significant increase in accuracy. It was concluded that 100 "price significant" items of equal value were optimal and would result in accuracy between 5 and 6%. The accuracy achievable for any estimating technique is thus dependent on the amount of detail which it contains.

3.11 Historical Cost Data

As described in chapter two, the cost data upon which estimates are based are inherently uncertain in nature and thus subject to variability. The effect of this variability on the accuracy of the estimate can be significant (Morrison, 1984). Stevens (1983, p.176), in her analysis of estimating accuracy, contends that forecasting errors are generally a result of the "lack of relevant price data". Morrison (1983) asserts that changing the manner in which cost data is selected and manipulated is the most likely means of improving estimating performance. He further maintains that the variability of the cost data as a whole can be determined by assessing the degree of variability of the mean of the lowest tenders. He thus suggests that the variability of price data is 5% (coefficient of variation).

In addition to the data's inherent variability, the degree of precision with which adjustments are made to it (to take into account differing project characteristics) will also affect the forecast's accuracy and will be determined by the expertise of the estimator.

Beeston (1975) claims that the reliability of the cost data will improve if it is obtained from several buildings rather than just one. Research conducted by Bennett, *et al.* (1981) to some extent confirms this. Table 3.1 below shows the level of increase in the accuracy of forecasts with greater cost data. It can be seen that the level of accuracy of the elemental estimate conducted from only one previous project improves by 40% (i.e. from 10% to 6% mean deviation and 13 to 7.5% cv) when a data-base of statistically analysed data is used.

Table 3.1 THE EFFECT OF THE DEGREE OF COST DATA USED ON THE ACCURACY OF ESTIMATING TECHNIQUES.
(Adapted from Bennett, *et al.* 1981)

ESTIMATING METHOD	MEAN DEVIATION (%)	CV (%)
1. Cost per square metre from one previous project	18	22.5
2. Cost per m2 derived by averaging rates from a number of previous projects	15.5	19
3. Elemental estimating based on rates from one previous project	10	13
4. Elemental estimating based on rates derived averaging the rates from a number of previous projects	9	11

5. Elemental estimating based on statistical analysis of all relevant data in the data base	6	7.5
6. Resource use and costs based on contractors estimating methods	5.5	6.5

Table 3.1 also indicates a relatively small increase in accuracy with the increase of the number of bills used where there is no statistical analysis of the data. Experiments conducted by Jupp and McMillan (1981) cited by Skitmore (1990), confirmed this. Their experiments were conducted with three quantity surveyors, who were required to price bills of quantities with an increasing number of previous bills from similar projects. The results showed that the bias of forecasts decreased "slightly" with an increase in amount of data (in the form of the bills), although no improvement was observed when more than two bills were used. However, no statistical significance has been attached to these results.

3.12 The Experience and Expertise of the Forecaster

The presence of uncertainty in the forecasting process necessitates subjective judgements to be made, which the estimator will base on his own interpretation of the project and the likely costs that it will incur. The ability or proficiency with which the forecaster makes such judgements can be expressed in terms of his/her experience and/or expertise.

Zahry (1982, p. 26) claims that "no matter how much measurement or description is available correct advice or a correct decision will not be forthcoming unless the information is interpreted by knowledge and experience." The forecaster's expertise in making cost significant decisions will thus not only affect the accuracy of the forecast, but also the success of the entire cost planning process in achieving its objectives.

Skitmore's (1985) research into the effect of expertise on the forecaster's performance showed that what distinguished "expert" quantity surveyors from "novices" was their ability to estimate the cost of a project with great accuracy when given no more information than the project type and size. The "novices" could only achieve such a level of accuracy when provided with a full bill of quantities to price.

Skitmore (1985) cited by Tan (1988) attributed the following characteristics to an "expert" forecaster (in order of importance) :

1. high recall ability,
2. self-professed expertise,
3. low mental imagery of physical characteristics of the building,
4. high general and specific estimating experience.

In addition, the surveyors with the greatest expertise were

thought to be more relaxed and confident, as well as being more concerned with maintaining familiarity with the market and overall price, (Skitmore 1985) cited by Skitmore *et al.* (1990). Property Services Agency (1980) observed accuracy to improve on certain types of projects with which the quantity surveyor had direct experience. Although Morrison (1984) maintains that accuracy will not automatically improve as a result of this experience, but rather by the means which knowledge and experience have been gained from previous projects and are related to future work. Thus for expertise to result from experience the estimator must "learn" from the errors made in previous projects. Ogunlana (1989, 1991) attributes the inability to "learn from experience" as largely a consequence of the estimator's failure to compare the estimate with the accepted bid on a regular basis. This comparison would provide feedback on their errors and help ensure that they are not perpetuated.

3.13 Other Factors

Other factors thought to affect the accuracy of price forecasts include the percentage of prime cost and provisional sums in the contract, plan shape (Skitmore, 1988) the gross floor area (Tan, 1988; Skitmore, 1988), tendering procedures (open, invited or negotiated tender, Wilson *et al.*, 1988), the type of contract (i.e. whether a bill of quantities was used, Wilson *et al.* 1988; Betts and Gunner, 1989; Gunner and Betts, 1990), the type of client (i.e. public or private sector), anticipated site conditions.

3.14 Conclusions

All of the above indicate factors that may influence the degree of accuracy with which the quantity surveyor forecasts the price of a project. However, most authors have failed to prove a uniform and statistically significant relationship between any of these factors and the level of accuracy.

CHAPTER FOUR : LITERATURE REVIEW OF OPINION AND EMPIRICAL STUDIES

CHAPTER FOUR

LITERATURE REVIEW OF OPINION AND EMPIRICAL STUDIES

4.1 Introduction

This chapter reviews much of the published literature dealing with the measurement of the accuracy of price forecasts and the factors affecting it. Two basic approaches to studying accuracy are used. The first, more traditional approach is the examination of retrospective records and existing projects and is most commonly used. The other approach is experimental in nature, involving the use of "dummy" projects.

The table below embodies the findings of research conducted over the last two decades dealing with the measurement of accuracy.

Table 4.1 ACCURACY OF DESIGN PRICE FORECASTS (Ogunlana, 1989)

SOURCE	MEAN ERROR	CV (%)	DATA SOURCE
Park (1972)	-	10-15	100 projects
Mitchell (1974)	+10	-	-
Beeston (1975)	-	7	large sample PSA projects
McCaffer (1975)	-	34	electrical services
McCaffer (1976)	+7.5	13.31	132 Belgian roads
Brown (1979)	+7.8	-	273 NASA projects

Brown (1979)	+7.49	16.77	22 projects
Kennaway (1979)	-30to+40	-	25 railway projects
Bowen (1980)	+9	-	attitude survey
Property Services Agency (1980)	-	13	958 public sector projects
McCaffery & McCaffer (1981)	-	6	15 school projects
Jupp & McMillan (1981)	-	18-24	experiment with 3 QS's and 9 projects
Thorpe (1982)	± 7	-	41 govt projects
Morrison (1983)	-	15	915 projects
Bennett, Morrison and Stevens (1981)	12	-	915 projects
Darko (1985)	-	12	33 projects
Skitmore (1987)	+10.51	-	
Tan (1988)	±26	13	100 USA govt projects
Ogunlana (1989)	-	12.77	51 projects

A chronological review of much of the published literature to date follows.

Beeston (1975) sought to determine a reasonable level of accuracy of price forecasts. He suggests that such a level of accuracy should be equivalent to the variability of the lowest tenders, since, quantity surveyors cannot realistically expect to improve their level of accuracy beyond that of the contractor. The variability of quantity surveyors' forecasts was established by comparing the estimated prices of individual items for "identical buildings". The coefficient of variation (CV) of the ratio of the lowest tender to the estimate was found to be 10.9%. Since identical buildings in practice cannot be found, adjustments were made to the CV to take into account the effect of different locations, building functions and contract values. The CV was reduced from 10.9% to 8.6%. This was then reduced to 7% for the "ability to allow for unidentified causes" (p.142).

The variability of the lowest tenders was suggested as being an average of 5.2%, which was increased to 6% to account for the uncertainty of the market. Thus Beeston estimates the quantity surveyors' accuracy in predicting the lowest tender to be 7%, which could only be improved to 6%, since any improvement beyond this would require an improvement in the contractors' estimating accuracy. The CV of 7% determined in this paper appears somewhat low in comparison with other authors (Morrison, 1983) 15.5%, Jupp and McMillan (1981) 18 -24%, McCaffer (1976) 13.13%, Property Services Agency (1980) 13%, Flanagan (1980) 15%, Ashworth and Skitmore (1983) 13 -18%). Its validity is questionable, since its derivation is highly theoretical in nature, the reliability of the adjustments to all the CVs' cannot be fully ascertained since there is a lack of supporting data, and such adjustments may not take into account the correlations between other factors affecting the variability of the estimate.

Property Services Agency (1980) As part of a larger, three year research project, data was collected from six different quantity surveying offices in the public sector. A total of 958 projects were used in an empirical study, which entailed a comparison of quantity surveyors' final pre-tender forecasts and the accepted tender figure. Each office's data was analyzed separately according to building type, project size and time to ascertain what effect these factors had on the level of accuracy. The accuracy was established by determining the mean error of the data. This was found to be approximately 13%.

It was found that accuracy tended to improve on large projects and types of projects where the quantity surveyor had more experience. Another factor influencing estimating performance was thought to be the method of estimating used. It was suggested that the major weakness of forecasting methods is that they are quantity related, while the contractors methods of estimating are both process and quantity related.

Bennett, Morrison and Stevens (1981) In a related study to that conducted in 1980, 915 projects from six private and one public quantity surveying office were collected. The percentage mean deviation was found to be approximately 9.8% (including P.C. and provisional sums), and 12% for 557 of the projects from which provisional and P.C. sums were removed.

Accuracy was found to improve where a large proportion of the work was of one particular type, which confirmed their previous conclusions that experience was positively correlated with accuracy. This improvement was attributed to a better understanding of the building type and a larger, more up-to-date store of relevant cost data. It was also concluded that accuracy improved by a small factor as the projects increased in value, which was assumed to be due to more effort and time being devoted to larger projects. Two main factors were suggested as causing the discrepancy between the quantity surveyors' level of accuracy (12%) and the contractors' (5 -6%), which were :

1. the addition of lump sums (for preliminaries; professional intuition and adjustments for inflation, region and specification level.)

2. the choice of items or elements to measure and price. It was shown that the accuracy could be improved to 5-6% by pricing only 100 cost significant items.

Ashworth and Skitmore (1983) In a review of much of the current literature relating to the accuracy of both quantity surveyors' and contractors' estimates, Ashworth and Skitmore suggest that Jupp's CV of 18 to 24% is the most reasonable assessment of accuracy at the detailed design stage. Property Services Agency (1980) approximate CV of 13%, which is substantiated by McCaffer's (1976) 13.13%, is considered to be the next most probable. They suggest a "suitable" accuracy for the early design stages to be 15% to 20% CV which would improve to 13% to 18% at detailed design. From these suggested levels of accuracy they conclude that an increase in the amount of design information available does not significantly increase the accuracy of the forecast, contradicting the assumption made by previous authors (Barnes, 1974); McCaffery, 1981).

Morrison (1983) analysed 915 projects at the detailed design stage. This study was based on the data obtained as part of the 1980/1 research programme. He calculated the average CV to be 15.45% (excluding P.C. and provisional sums). This variability was attributed to :

1. the variability of tenders (6.6%)
2. the use of insufficient cost data (5%)
3. the inherent variability of the forecasting technique (1.85%)
4. the variability of the adjustments made to the cost data (6.9%)
5. the use of unsuitable cost data (11%).

However these figures are of questionable reliability since they are not confirmed by any supporting evidence and are merely a "best guess".

Stevens (1983) Based on 1013 projects from seven different offices an analysis of accuracy established a CV of 13%. The reasons for the errors were largely ascribed to the lack of relevant cost data, which coincides with Morrison's (1983) conclusions.

Flanagan and Norman (1983) examined the forecasting performance of two public sector quantity surveying departments on 166 projects dating from 1971 to 1978. Although most of the projects undertaken by these departments were schools, (and thus the surveyors would have greater experience with this type of project) a significantly better level of accuracy was not shown. Regression analysis was performed to determine if any pattern occurred in the errors recorded, but no consistency was shown to occur. The first department (Council A) overestimated by 11.5% on average, while Council B overestimated on the small projects but underestimated on large. The data also showed that the degree of error tended to decrease as the number of bidders increased. In addition, from 1971 to 1974 the estimates were consistently underestimated which the authors attributed to the slow response of the Councils to the changing market conditions. A feedback mechanism was proposed which, by monitoring the estimating performance, could have detected and remedied the problem.

Skitmore (1986) considers the importance of the expertise of the forecaster in predicting the tender. From the literature studied he concluded that an expert surveyor could solve a more complex problem faster and with greater accuracy than a novice. An experiment was conducted in 1984 with twelve practising quantity surveyors to determine the effect that the amount of design information available has on the surveyor's ability to accurately predict the contractor's estimate. Each surveyor performed two estimates chosen from a selection of five recent projects. It was found that the projects were overestimated by an average of 10.51%. The results showed that the level of accuracy increased significantly once the information was received and only negligibly when the drawings were received. It appeared that, what distinguished the experts from the novices, was the quality of the first estimate which was based on only the project size and type.

Skitmore (1987), based on the research conducted in 1984 (Skitmore, 1986), attempted to determine the effect that the amount of design information available has on the surveyors' ability to accurately predict the contractor's estimate. This study challenged the accepted assumption that forecasting accuracy significantly improves as more design information becomes available. This was first noted by Ashworth and Skitmore (1983) from their review of published literature. The research showed that the accuracy of the self-professed "experts'" forecasts did not significantly improve with the increased amount of design information. Given only the building type and size, an "expert" was able to predict the tender figure with an equivalent degree of accuracy as the average surveyor at detailed design stage. However, the sample size was extremely small and thus insufficient to draw any firm conclusions from. Also, the information types used may not have been entirely appropriate. Since the projects did not contain any unusual features, Skitmore suggests that the tests may have been too "easy" for the experts. Other reservations, Skitmore contends, include the accuracy of the accepted tender against which the forecasts were compared and the reliability of the methods used to adjust for the effects of project type.

Wilson, Sharpe and Kenley (1987) sought to determine which factors affect forecasting accuracy. The effects of eight given factors were measured by establishing the percentage of projects which were re-submitted for further approval, re-design, or negotiation, due to the fact that the forecast exceeded the tenders by more than 10%, where each factor prevailed. This, in effect, measured, for each factor, the proportion of projects on which the quantity surveyor's forecast differed by 10% or more from the accepted tender. Data were obtained for 410 Australian PWD projects from 1979 to 1982 and were analyzed using logistic regression. From the factors which were analyzed, only three appeared to significantly affect the estimating accuracy. It was found that the size (i.e. value) of a project was positively correlated with re-submission, thus the higher the value of the project the greater the probability that the error of the forecast would exceed 10%. However, Skitmore (1988) claims that these figures "offer little support" for the suspected correlation between project size and accuracy. Wilson *et al.* further concluded that where Bills of Quantities were used the rate of projects re-submitted was significantly lower. In addition they found that the greater the number of tenderers, the lower the rate of re-submissions.

Runeson (1988) statistically analyzed the distribution of tenders on Australian PWD and DHC (Department of Housing and Construction) building contracts. He established the CV of successful tenders to be 4.9%, which is marginally lower than those stated by other authors which fall in the range of 5 to 7%. Runeson suggests that this slightly lower variation is due to the separation of the projects according to the number of tenderers.

Skitmore (1988) examined two samples of data, the first containing 67 American government building and engineering projects, and the second, 33 British government building projects. The mean accuracy for the American projects was found to be -12.38% (21.53 standard deviation) while for the British it was determined to be -4.91% (17.22 standard deviation). The high standard deviation of both samples showed a poor consistency in estimating. Although statistically the analysis did not show any correlation between the factors examined and accuracy, visual scrutiny proved to confirm the findings of previous authors, which were reviewed in the paper. The lack of statistical confirmation of such findings was largely attributed to the extremely small sample sizes of projects.

Betts and Gunner (1989) measured the accuracy of 46 Singapore building projects, 16 of which were tendered on the basis of specification and drawings. This sample size is extremely small, which causes the outcome to be statistically unreliable. The coefficient of variation of the ratio of the forecast to the lowest tender for the sample was found to be 9%. Analysis also showed that quantity surveyors generally tended more often to overestimate costs and that the level of accuracy improved with the increase in value of the projects. However, it should be noted that this data was obtained from a single quantity surveying organisation and the results and conclusions are thus applicable only to that organisation. Thus the results cannot be assumed to represent a general trend applicable to all quantity surveyors.

Ogunlana (1989) conducted an opinion and empirical survey of 51 projects from seven County Council Road and Transport Departments in the U.K.. The percentage mean error of these projects was found to be 12.77%, the mean absolute error to be 16.44% using the lowest bid as the datum against which the accuracy was measured. The empirical survey showed that the individual expertise of a surveyor or office (where more than one surveyor in the office is involved in estimating) appears to significantly affect the accuracy of the forecast. This confirms those views stated in the opinion survey. The only other factor for which its correlation with accuracy appeared to be statistically significant was project location. The analysis suggested that, although the prices did not differ within the compact regions of each county council, they did however vary significantly between regions. Although the size of the project showed a positive relationship with accuracy, it was not statistically significantly. Five factors were re-analysed (i.e. project value, geographical location, number of bidders, year of tender and the office of origin) and measured against the datum which the office claimed that it aimed at predicting. Using these parameters as the measurement datum, none of the factors showed any statistically significant relationship with the level of accuracy achieved.

CHAPTER FIVE : EMPIRICAL AND OPINION SURVEY

CHAPTER FIVE

OPINION AND EMPIRICAL SURVEY

5.1 Introduction

An opinion survey was initially undertaken to (i) determine the levels of accuracy that professional quantity surveyors perceive they attain and to (ii) ascertain which factors they believe affect their estimating performance. An empirical study was then conducted to ascertain the actual levels of accuracy achieved and the factors which influence it. To substantiate the problem, a comparison of the actual and perceived levels of accuracy was carried out, determining whether quantity surveyors are in fact unaware of their levels of accuracy achieved. Once the actual levels had been established, the hypothesis (as stated in Chapter One) that it is possible to determine which factors were associated with higher level of accuracy then others, could be validated.

5.2 The Opinion Survey

5.2.1 Introduction

A questionnaire was randomly distributed to professional quantity surveying offices in Cape Town. Thirty responses from were recorded from the total of sixty questionnaires which were distributed. (A sample of which can be found in the appendix.)

The aim of this questionnaire was largely to determine the levels of estimating accuracy that quantity surveyors perceive they attain. In addition, questions were asked relating to the estimating methods utilised, type of cost data used, levels of design information available at each stage of the design process, feedback of accuracy, as well as the factors thought to influence accuracy.

5.2.2 The Results

5.2.2.1 The Level of Accuracy Perceived to be Attained

For each stage of the design process the quantity surveyors asked what they perceived their level of accuracy to be. The following table shows the percentage of respondents which perceived their level of accuracy to fall within the stated accuracy category at the respective design stage.

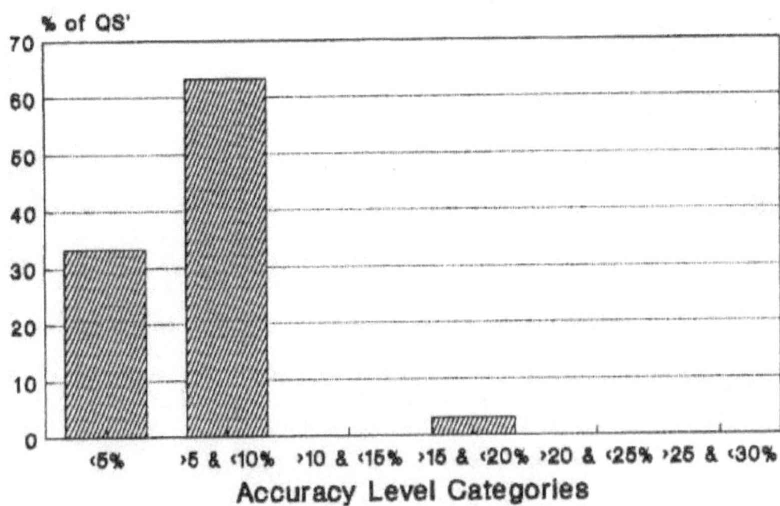

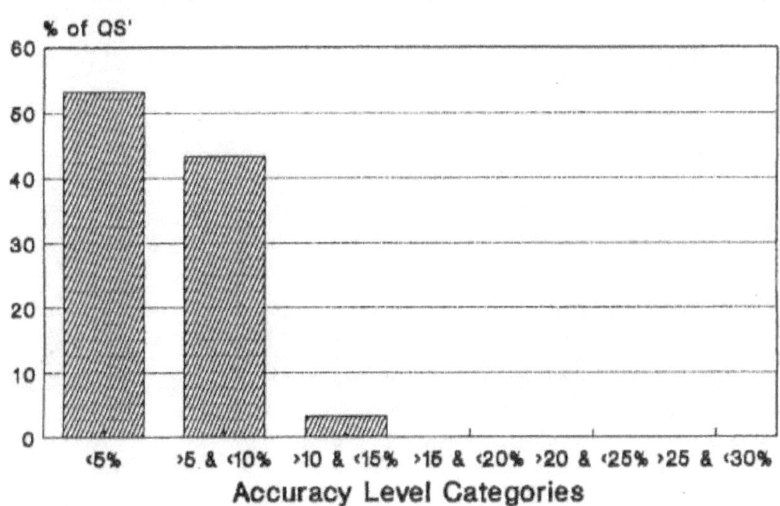

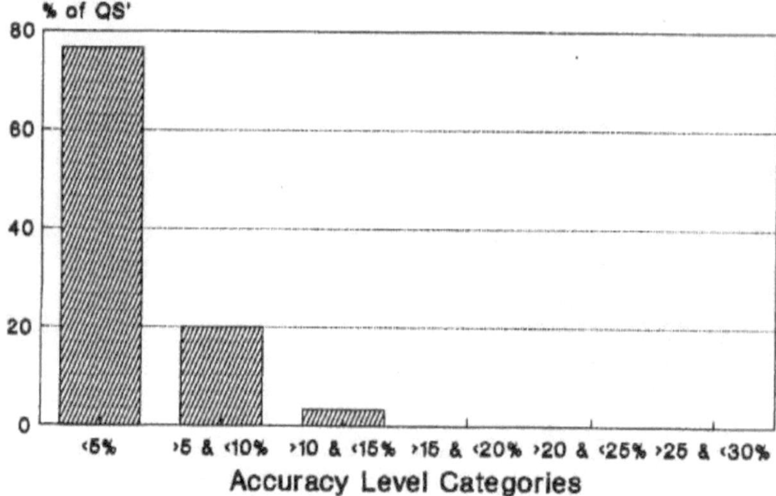

TABLE 5.1 PERCEIVED LEVELS OF ACCURACY

STAGE OF DESIGN	LEVEL OF ACCURACY						
	<5%	>5 & <10%	>10 & <15%	>15 & <20%	>20 & <25%	>25 & <30%	Average
Inception	10.3	44.8	41.4	0	0	3.4	12.24
Feasibility	24.1	41.4	31.4	3.4	0	0	10.67
Sketch Design	33.3	63.3	0	3.4	0	0	8.67
Detail Design	53.3	43.3	3.3	0	0	0	7.50
Tender	76.7	20.0	3.3	0	0	0	6.33

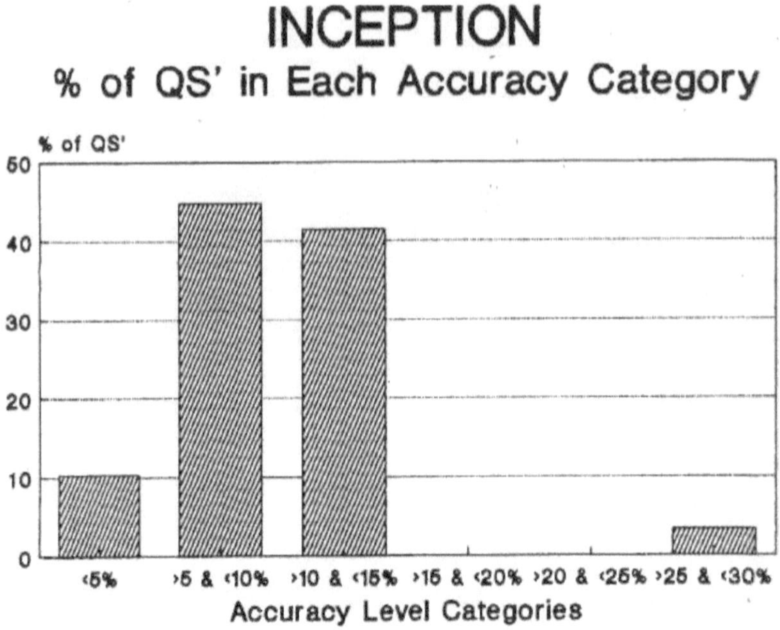

Figure 5.1

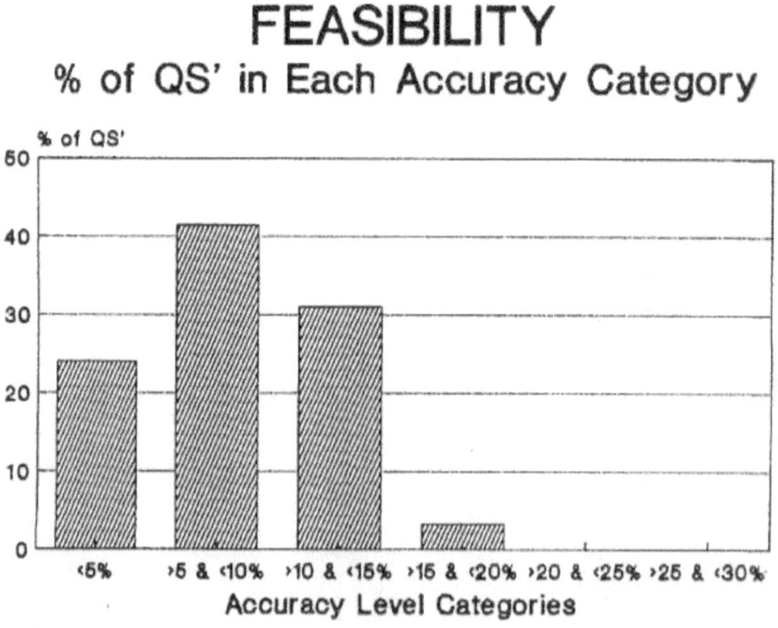

Figure 5.2

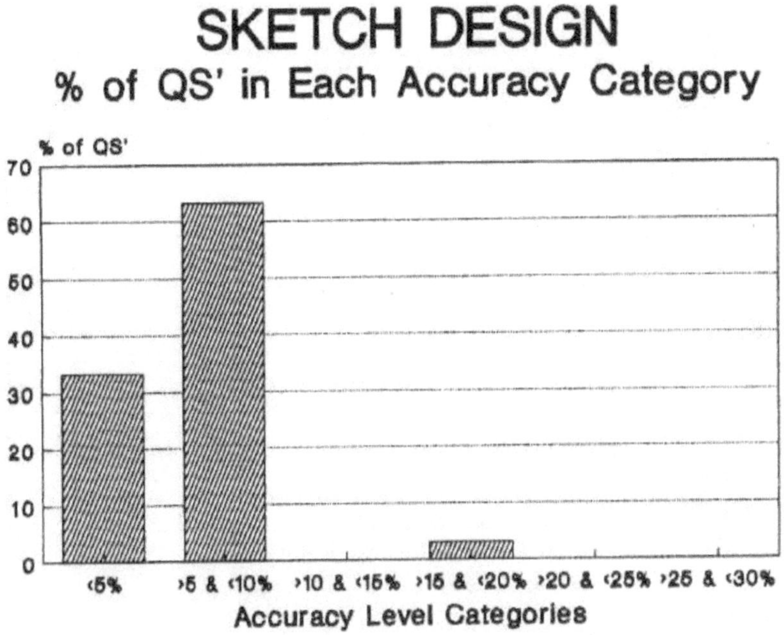

Figure 5.3

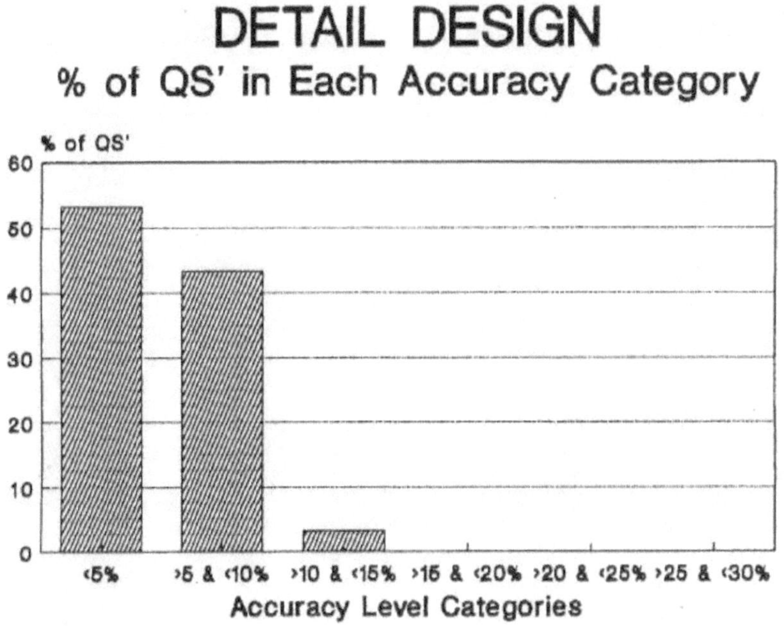

Figure 5.4

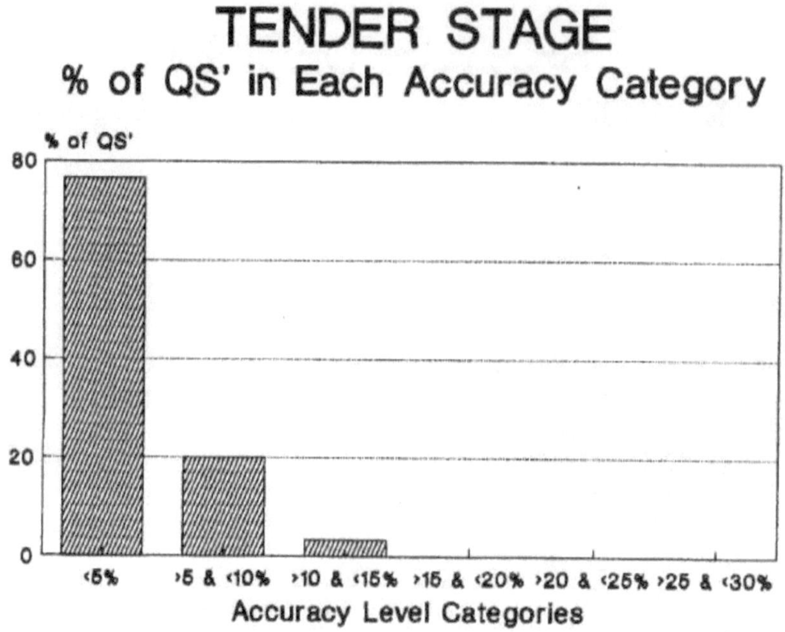

Figure 5.5

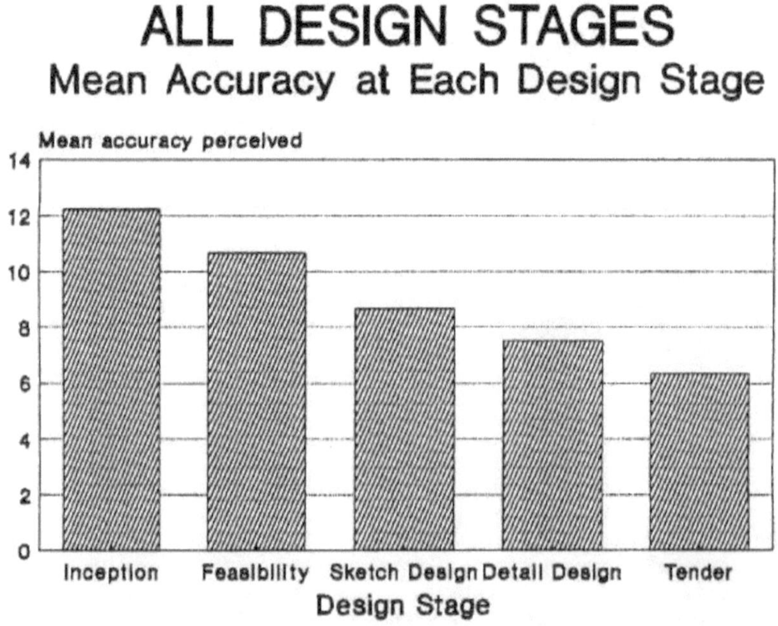

Figure 5.6

From the inception stage of the project to sketch design the majority of the respondents perceive their forecasts to be within five to ten percent of the tender figure. This figure is then perceived (by the largest portion of the respondents) to improve to within five percent. The average percentage accuracy improves from 12.24% to 6.33% from inception to tender.

5.2.2.2 The Relationship between Design Stage and the Estimating Techniques Used

When questioned which method of estimating was most often used at each stage of the design process the responses were as follows :

TABLE 5.2 DESIGN STAGE VERSUS FORECASTING METHODS USED

	% USE OF METHOD					
DESIGN STAGE	M2	ELEM	AQ	AQ/ ELEM	PWD	BOQ
Inception	67	17	13	3	-	-
Feasibility	17	50	30	3	**	-
Sketch Design	-	50	30	20	**	-
Detail Design	-	43	37	10	**	10
Tender	-	7	-	-	-	93

KEY TO TABLE 5.2 :

M2 = Superficial area method, i.e. price expressed per square metre of the building

ELEM = Elemental cost estimate

AQ = Approximate or rough quantities

AQ/ELEM = Either the Approximate quantities or the elemental method was used, the respondent claimed that the methods were used with equal frequency at these stages.

PWD = Storey enclosure method, i.e. that devised by the Public Works Department (i.e. PWD)

** = At these stages one respondent claimed to use either the PWD system or the approximate quantities or elemental estimating methods.

Table 5.2 on the previous page shows that at inception the superficial area method is the most commonly used estimating technique. This is most probably as a result of the lack of design information available at this time which precludes the use of more detailed techniques without making substantial assumptions regarding the design.

It appears from the table that as the design progresses, the estimating method used increases in complexity and detail. From Feasibility to Detail Design elemental estimating is the most frequently used technique. At the Tender stage the quantity surveyor will almost always price his completed bills of quantities.

5.2.2.3 Type of Cost Data Used

At each stage of the design process the type of cost data used was stated. The table on the following page shows the percentage use of each type at each stage of the design process.

TABLE 5.3 RELATIONSHIP BETWEEN DESIGN STAGE AND COST DATA

DESIGN STAGE	% USE OF TYPE OF COST DATA USED In-house only	 In-house & quotes	 In-house & PD	 In-house, quotes, PD
Inception	90	10	-	-
Feasibility	54	43	-	3
Sketch Design	40	50	3	7
Detail Design	30	64	3	3
Tender	33	60	-	7

KEY TO TABLE 5.3 :

In-house = Data stored and processed within the office, assimilated from projects which employees were directly involved with.

Quotes = Quotes obtained from subcontractors and suppliers.

PD = Published data in the form of rates and prices, obtained from journals or price books, i.e. this data is not generated from within the office. This data does not include the economic indices (such as the BER index, etc.) used to escalate costs to take into account the inflation of prices.

It an be seen from the above table that very little, if any, published data is used from estimating purposes. South Africa, in contrast with the U.K., has very little published data available. These findings support those of Morrison and Stevens (1980), whose study also showed that published data was the least popular source from which cost data is obtained. The table also indicates that the use of specialist quotes increases significantly as the design progresses and more detailed information becomes available.

The surveyors were also asked in which form they stored this data. Their replies were as follows :

Cost Analyses and Updated Bills = 57%

Updated Bills Only = 20%

Cost Analyses Only = 20%

Worked-up Rates = 3%

5.2.2.4 The Relationship between the Amount of Design Information Available and Accuracy Achieved

The surveyors were then questioned what design information was generally available at each stage, what they perceived their level of accuracy to be at each of these stages and whether they believed that their level of accuracy "improves significantly" once detailed design information becomes available and why.

From the responses it was found that at inception a rough sketch or line drawing is usually provided, occasionally with a brief specification. This drawing is then amended and updated at feasibility, with only a minor addition of further detail. By sketch design the plans, sections and elevations are often available with a brief specification. Usually by detail design the working drawings are practically completed. By tender it appears normal for fully annotated working drawings to be supplied.

After establishing what degree of design information is generally available, the surveyors were then questioned whether they thought their accuracy "significantly improves" as more design information becomes available. Skitmore (1987) contends that the amount of design information available will not significantly improve the accuracy of an "expert" quantity surveyors' forecast. 76.7% of the quantity surveyors believed their accuracy did significantly improve. All concurred that the increase was due to the greater detail available resulting in a better insight into the project. The remaining 23.3% which concluded that accuracy should not improve unanimously agreed that the inclusion of contingency sums for expected costs would insure a consistent level of accuracy throughout the design process.

It was then asked whether they believed there was sufficient design information available at the inception of the project to perform a "realistic" forecast. 40% considered there was, 53.3% thought there wasn't, while the remaining 6.7% thought that it was possible sometimes. Of those that replied that it was not possible the majority (56.25%) thought that a "realistic" estimate could be performed only at the sketch design stage.

5.2.2.5 The Relationship between Accuracy and the Type of Project

When questioned whether a significantly higher the level of accuracy is achieved on certain types of projects (at tender stage), 58.6% believed their level of accuracy to be higher, while 13.8% were unsure and the remaining 27.6% did not believe their accuracy to improve. Those who considered their accuracy to increase were then asked what level of accuracy they attained. All responded by stating the same level of accuracy attained for normally at the tender stage, which contradicted their previous statement.

The 58.6% of the respondents who believed their level of accuracy to be greater on certain types of projects stated the following types :
(Note : the percentages below indicate the percentage of respondents who believed that their accuracy was greater on these types)

Commercial	= 68.75 %	
Schools	= 43.75 %	
Flats	= 12.50%	Total Residential = 31.25%
Luxury Homes	= 6.25 %	
Homes	= 6.25%	
Housing Schemes	= 6.25%	
Gyms	= 6.25%	
Industrial	= 6.25%	
"If familiar"	= 12.50%	

When questioned why a higher level of accuracy was perceived on these types of projects the following reasons were given :

Experience	= 62.5%
Repetition	= 25%
Predictable	= 6.25%
More Data Available	= 6.25%

The above indicates that it is perceived that experience is a significant factor in determining the level of accuracy achieved. When combined with experience, it is believed that the type of project influences accuracy.

When questioned directly whether experience was important in the forecasting process it was found that it was considered :

"Not important" by 3.3%

"Significant" by 3.3%

"Important" by 0%

"Very important" by 43.3%

"Essential" by 50% of the respondents.

5.2.2.6 Factors Perceived to Affect Accuracy

The surveyors were asked to what extent a given set of factors influence the accuracy of their forecasts. This was done by rating the factors on a scale of 1 to 5, one being no influence and five being essential to the forecast's accuracy. The table below indicates degree of importance the respondents attached to each factor. The percentages indicate the percentage of surveyors which rated the factor at that level.

TABLE 5.4 FACTORS PERCEIVED TO INFLUENCE ACCURACY

FACTOR	EXTENT OF INFLUENCE				
	1	2	3	4	5
Type of Project	10.0	6.7	**40.0**	33.3	10.0
Design Information Available	3.3	6.7	3.3	30.0	**56.7**
Cost Data Available	0.0	10.0	26.7	**50.0**	13.3
Experience	0.0	6.7	13.3	**66.7**	13.3
Expertise	3.3	0.0	10.0	40.0	**46.7**
Project Location	6.7	**33.3**	30.0	20.0	10.0
Project Value	16.7	**40.0**	36.7	3.3	3.3
Degree of Services	3.3	23.3	**46.7**	23.3	3.3
Project Complexity	0.0	20.0	33.3	**40.0**	6.7

Present Market Conditions	3.3	10.0	20.0	26.7	**40.0**
Expected Future Market Conditions	10.0	13.3	23.3	**36.7**	16.7
Project Duration	3.3	10.0	**56.7**	23.3	6.7
Cost Limits	20.7	**31.0**	20.7	17.2	10.3
Site Conditions	0.0	26.7	**30.0**	26.7	16.7
Design Team	6.7	20.0	23.3	**36.7**	13.3

Highly Rated Factors :

Design information available, expertise, present market conditions.

Moderately Rated Factors :

Type of project, future expected market conditions, site conditions, design team, project duration, project complexity, degree of services, cost data available, experience.

Lowly Rated Factors :

Cost limits, project value, project location.

These factors are perceived by the respondents to exert a varying degree of influence over the accuracy. The actual effect of many of these factors will be tested in the empirical study.

5.2.2.7 The Perceived Expertise of the Respondents

The respondents were asked to evaluate their own estimating ability by rating different factors on a scale of one to five. One being no ability at all, while five indicates exceptional ability. The table below shows the percentage of respondents who perceive their ability to fall within the stated category.

TABLE 5.5 PERCEIVED LEVELS OF ABILITY

ABILITY TO :	EXTENT OF ABILITY 1	2	3	4	5
Identify Significant Cost Aspects	0.0	3.3	20.0	**56.7**	20.0
Cope with Insufficient Design Detail	0.0	0.0	26.7	**53.3**	20.0
To Visualise the Building	0.0	0.0	23.3	**60.0**	16.7
Intuition	0.0	3.3	30.0	**63.3**	3.3
Experience	0.0	0.0	26.7	**63.3**	10.0
Understanding and Knowledge of Market Conditions	0.0	3.3	**60.0**	33.3	3.3
Analytical Ability	0.0	3.3	33.3	**56.7**	6.7
Logical & Systematic Approach	0.0	0.0	20.7	**65.5**	13.8
Memory of Similar Project Details	0.0	0.0	36.7	**56.7**	6.7
Personality Factors	0.0	0.0	**57.1**	35.7	7.1
Length of Time Spent in Profession	0.0	3.3	26.7	**56.7**	13.3
Qualifications	0.0	10.0	36.7	**46.7**	6.7

Most of the respondents considered themselves to have "significant" ability (i.e. scale 4) with respect to almost all the factors.

5.2.3 Conclusions

From the above questionnaire it can be concluded that :

1. The perceived level of accuracy improves from 12.24% to 6.33% (mean percentage error) over the design process.

2. The elemental system of estimating is the most popular forecasting technique at all stages, except inception and tender.

3. Initially only "in-house" cost data is used for estimating. As the design progresses quotes combined with the in-house data are predominantly use. Published data is seldom utilised. The "in-house" data is usually stored in the form of cost analyses as well as old priced bills.

4. The majority of the respondents (53%) believe that a "realistic" estimate cannot be performed at inception.

5. 76.6% perceive their level of accuracy to improve significantly as the design evolves. The remaining 23.4% consider the inclusion of contingency sums to ensure a consistent level of accuracy throughout the design process.

6. The type of project is considered by 58.6% to influence the

accuracy achieved. The effect of the project type is believed to result from the experience with such projects. Commercial projects are associated with the highest level achieved.

7. Experience is considered to be "essential" (by 50%) or "very important" (by 43.3%) to the forecast's accuracy.

8. Design information available, expertise, and the prevailing market conditions are thought to be the most significant factors which influence accuracy.

5.3 The Empirical Study

5.3.1 Introduction

Data was obtained from the Master Builders Association (MBA) pertaining to the projects tendered over the last ten years. From these records the project name, type, location, and tender figures were extracted for each project. The same quantity surveying offices which responded to the opinion survey were then approached to furnish their forecasted prices for the projects on which they were involved. In total 243 projects were analysed, all of which were obtained from and measured by private quantity surveying firms.

In addition, 45 public sector projects measured and obtained from the Public Works Department (PWD) were also analysed.

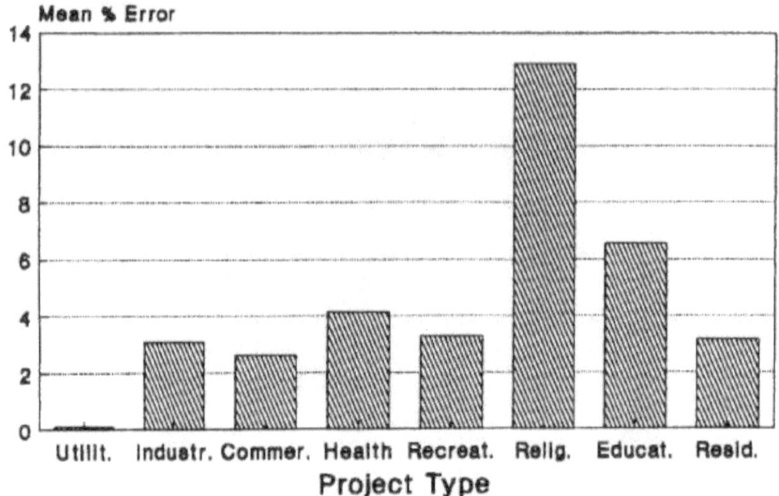

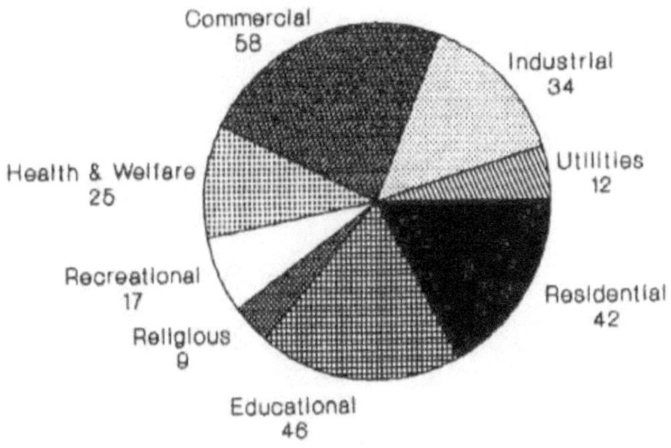

5.3.2 The Analyses

The accuracy of the estimates was measured in terms of mean percentage error and the mean absolute error. These figures were calculated both net and gross (i.e. with and without allowances for provisional and prime cost sums). Since the fixed sum allowances, as their name suggests, remain constant, they are not required to be estimated by the contractor. The inclusion of these allowances will increase the contract sum, and since accuracy is measured in relation to the contract sum, the level of accuracy will appear deceptively high. For this reason they have not been included in the accuracy calculations and thus all measures of accuracy are net. The variability of the contractors' bids is measured by means of the coefficient of variation.

Accuracy has been measured in terms of the mean percentage error and the mean deviation. The mean percentage error, because it takes signage into account, allows the errors to cancel each other out (i.e. a positive error cancelling out a negative error). The resultant figure will thus appear misleadingly low, not giving any indication of the total spread of the deviations that make up this average value. The mean deviation, because it is measured in absolute terms, will reveal the actual magnitude of forecasting error. For this reason, in the analysis below, the mean deviation is preferred as a measure of accuracy, although both figures are always stated.

This section aims to analyze as many of the factors discussed in chapter three as possible. However the effects of some of these factors proved impossible to examine as a result of the nature of the data received. Such factors included the project complexity, duration, the level of design information available, the historical cost data available and the experience of the forecaster. For the reasons discussed in chapter two, the data will only compare the final estimated figure with the lowest tender figure.

5.3.3 Data Estimated by Private Quantity Surveying Offices

This data was analysed in terms of the following criteria :

1. The value (i.e. size) of the project
2. The type of contract
3. The number of bids
4. The variability of the bids in terms of percentage tender range and the coefficient of variation
5. The geographical location of the project
6. The percentage of fixed sum allowances
7. Alteration/renovation and new work
8. The year of tender (i.e. state of the market)
9. The quantity surveying office
10. Private and public sector projects
11. The performance of MBA members

The analysis of the 243 projects showed that the accuracy of quantity surveyors' forecasts was 11.78% (mean net deviation), while the mean percentage error was 4.52%. The variation of the tenders, measured in terms of coefficient of variation and percentage bid range, was found to be 5.44% and 18.38%, respectively.

5.3.3.1 The Value/Size of the Project

The projects were divided into size categories of :
(1) not exceeding one million rand, (2) between one and two million, (3) between two and four million, (4) between four and eight million, (5) between eight and sixteen million, and those projects (6) exceeding eight million rand in value. The pie chart on the following page shows the breakdown the number of projects over each price category. In addition, the mean deviations (i.e. the mean absolute percentage errors) and the mean percentage errors of the forecasts and the coefficient of variation (CV) of the tenders for each price category are indicated on the bar charts. These charts are summarised in the accompanying table.

CONTRACT VALUES
Breakdown of Projects in Size Categories

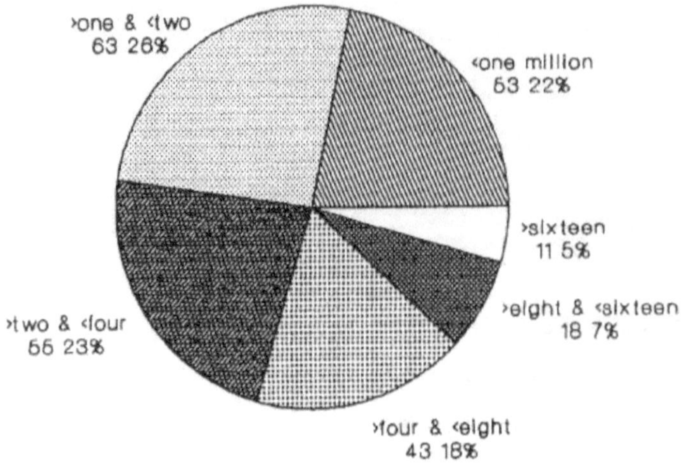

Figure 5.7

Mean Deviation for Each Value Category

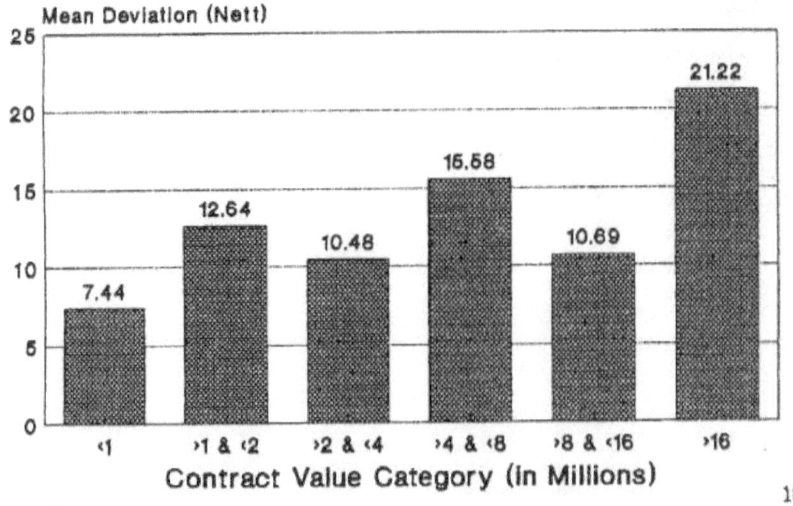

Figure 5.8

CONTRACT VALUES

Mean % Error for Each Value Category

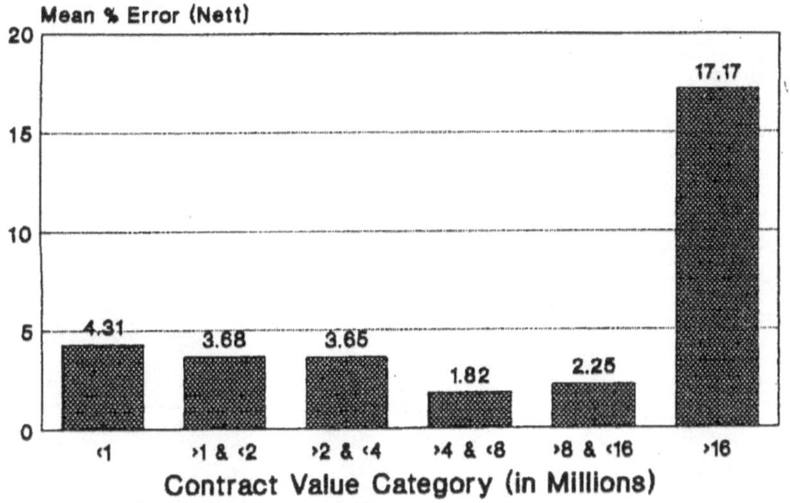

Figure 5.9

CV's of Tenders for Each Value Category

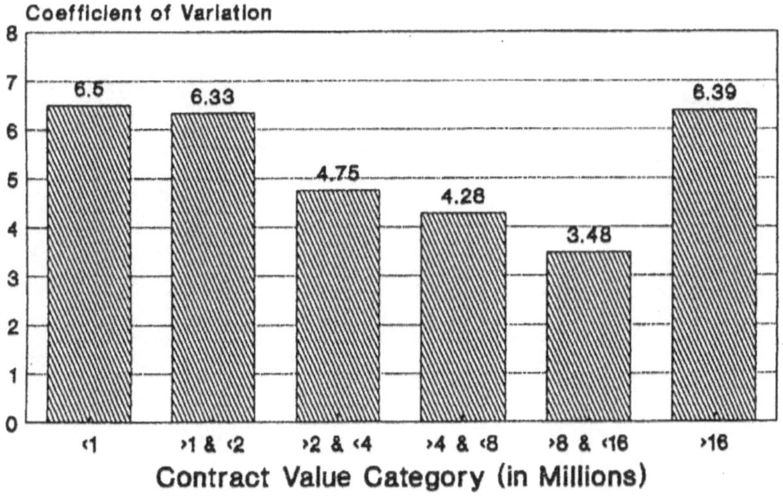

Figure 5.10

TABLE 5.6 ANALYSIS ACCORDING TO PROJECT VALUE

Project value (R millions)	No. of Projects	Mean Deviation	Mean % Error	CV of Tenders
< 1	53	7.44	4.31	6.50
>1 & <2	63	12.64	3.68	6.33
>2 & <4	55	10.48	3.65	4.75
>4 & <8	43	15.58	1.82	4.28
>8 & <16	18	10.69	2.25	3.48
> 16	11	21.22	17.17	6.39

The mean percentage error appeared to be consistent over all categories, with the exception of the sixteen million rand or greater value category. This category also showed a significantly higher mean deviation. The CV's indicated the variation of the bids to relatively low and consistent.

5.3.3.2 The Type of Project

The projects were divided into and analysed in terms of the CI/SFB work groups. Table 5.7 shows the results of this analysis.

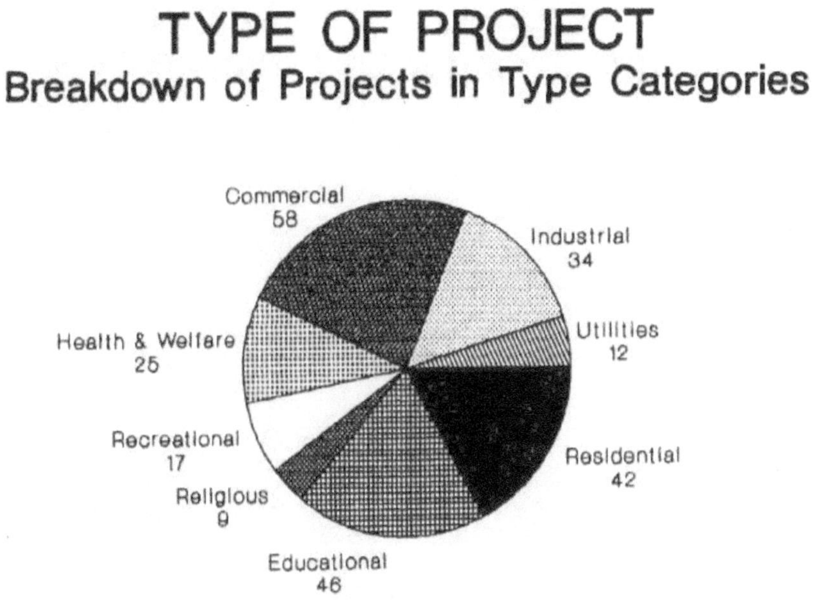

Figure 5.11

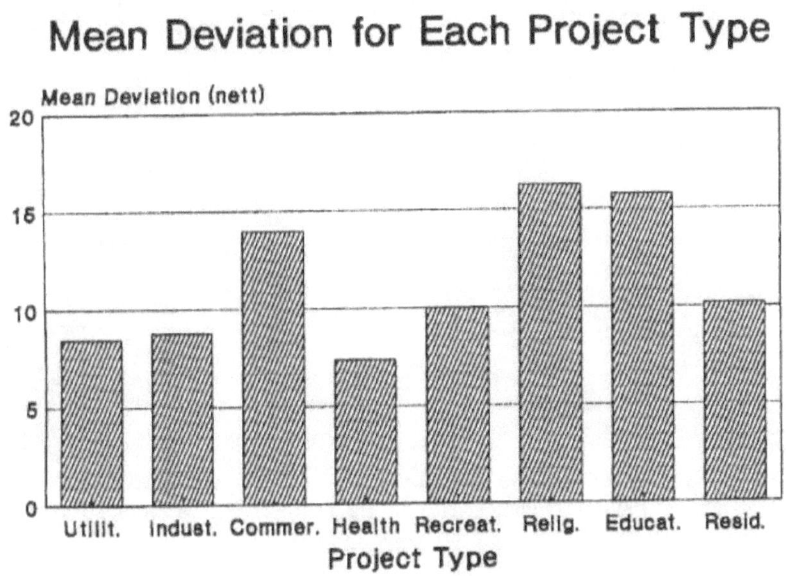

Figure 5.12

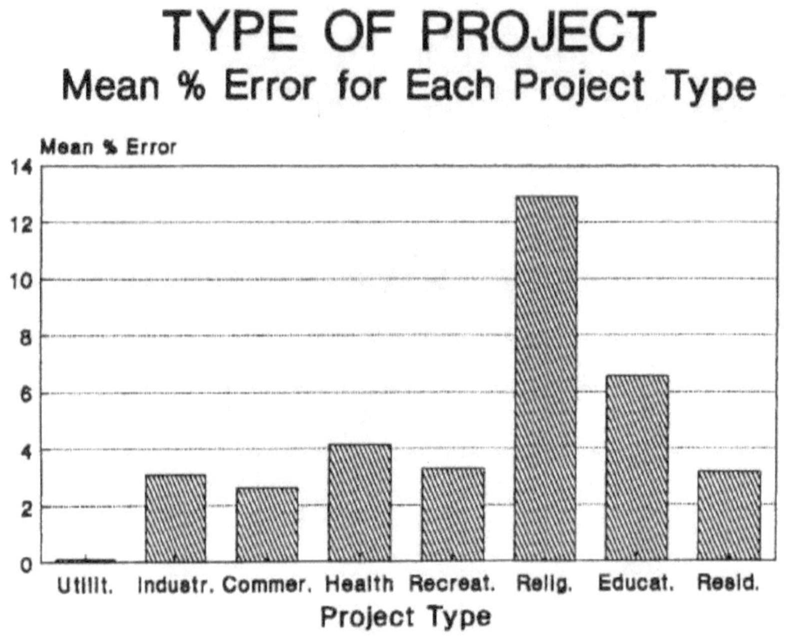

Figure 5.13

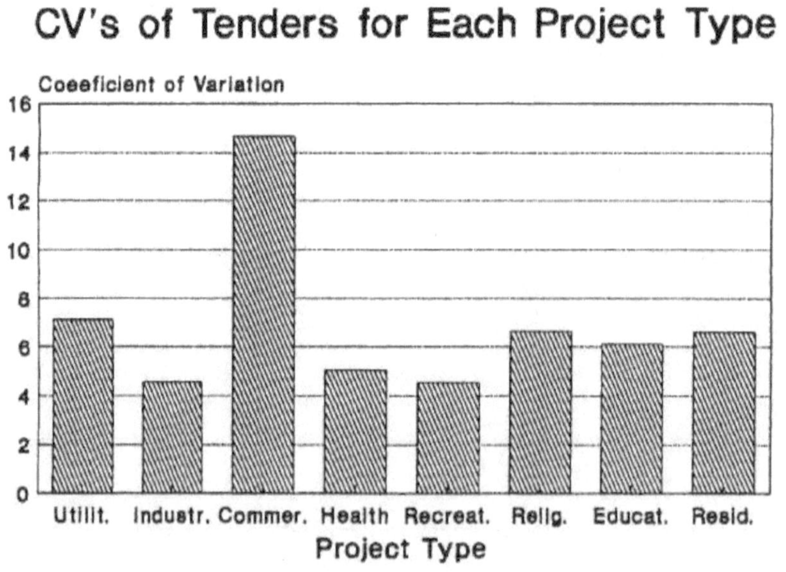

Figure 5.14

TABLE 5.7 ANALYSIS ACCORDING TO THE TYPE OF PROJECT

Project Type	No. of Projects	Mean Deviation	Mean % Error	CV of Tenders
Utilities	12	8.47	0.13	7.12
Industrial	34	8.78	3.08	4.56
Commercial	58	13.95	2.62	4.47
Health, Welfare	25	7.36	4.14	5.07
Recreational	17	9.97	3.29	4.53
Religious	9	16.29	12.92	6.62
Educational	46	15.76	6.58	6.13
Residential	42	10.19	3.17	6.59

The table above show religious and educational projects to be on average the least accurately estimated of all types of projects. The consistency of the CV's indicate that the variability with which the contractors estimate is not affected by the type of project.

5.3.3.3 The Number of Bids

The projects were grouped according to the number of bids received for each contract. These groups were less than 5 bids, exceeding five and not exceeding ten bids, and exceeding ten bids. The four graphs on the following pages illustrate the mean percentage error, mean deviation and CV, and are summarised in the following table.

NUMBER OF BIDS
Number of Projects in Each Bid Category

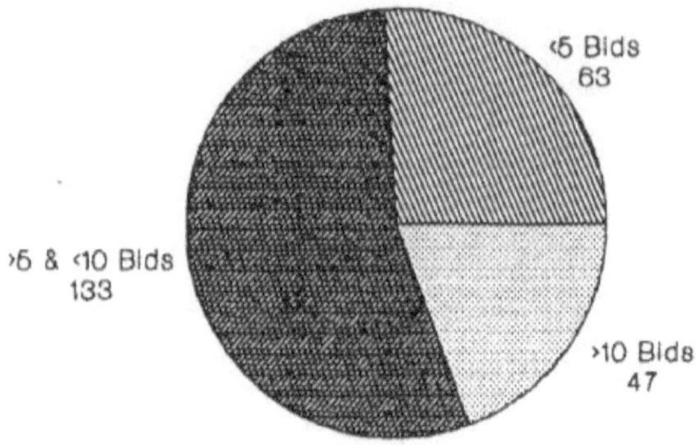

Figure 5.15

Mean Deviation for Each Bid Category

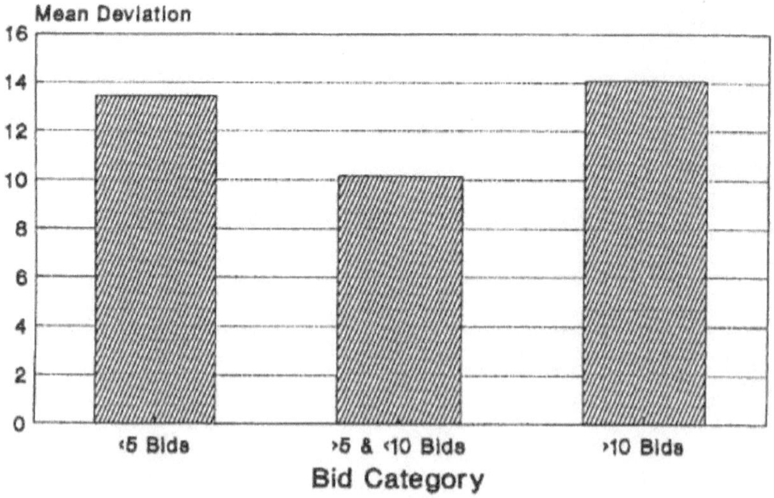

Figure 5.16

NUMBER OF BIDS

Mean % Error for Each Bid Category

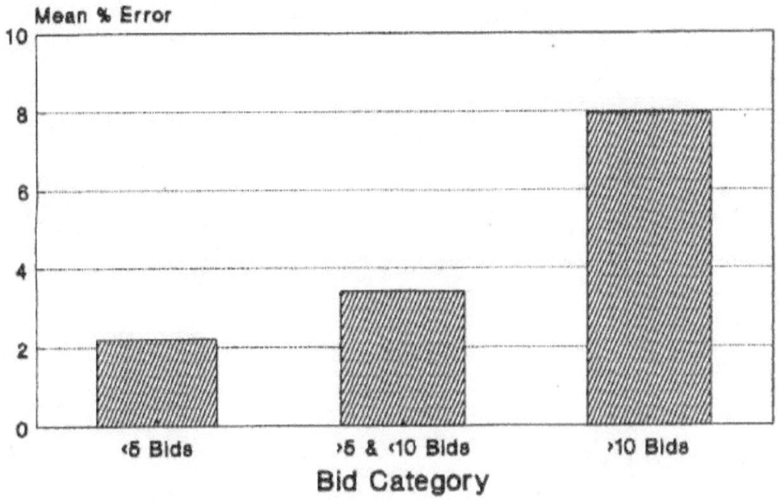

Figure 5.17

CV of Tenders for Each Bid Category

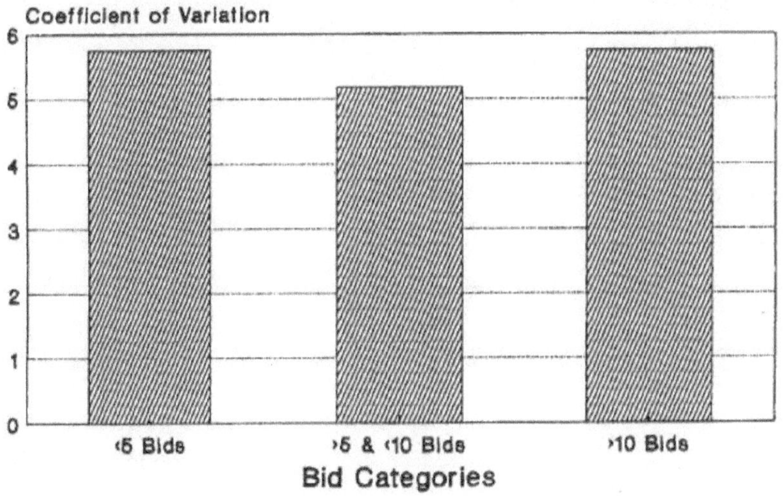

Figure 5.18

TABLE 5.8 ANALYSIS ACCORDING TO THE NUMBER OF BIDS

Number of Bids	No. of Projects	Mean Deviation	Mean % Error	CV of Tenders
< 5	63	13.44	2.22	5.76
>5 & <10	133	10.18	3.42	5.17
> 10	47	14.08	7.95	5.76

The mean deviation shows the level of accuracy to remain relatively consistent for each bid category, which suggests that the number of bidders for a contract does not affect its accuracy. In addition, the similarity of the CVs shows that the number of bidders tendering for the contracts did not affect the tender variability. This appears to indicate that the variation of tenders is not necessarily a function of the number of tenderers. The relatively high positive mean percentage error for the >10 bidders category indicates that the quantity surveyors have tended more often to overestimate the contract sum.

5.3.3.4 The Variability of the Bids

The variability of the contractors tenders was measured in terms of the coefficient of variation and the percentage tender range.

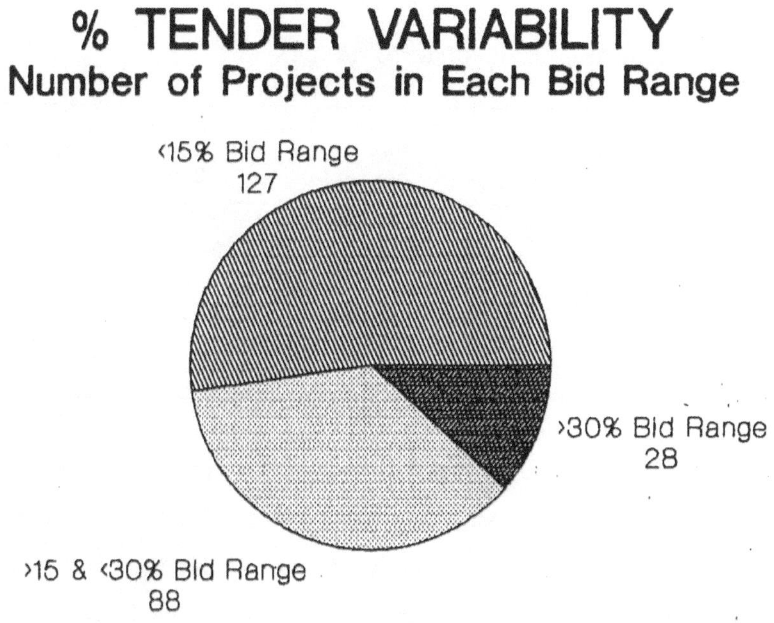
% TENDER VARIABILITY
Number of Projects in Each Bid Range
<15% Bid Range
127
>30% Bid Range
28
>15 & <30% Bid Range
88

Figure 5.19

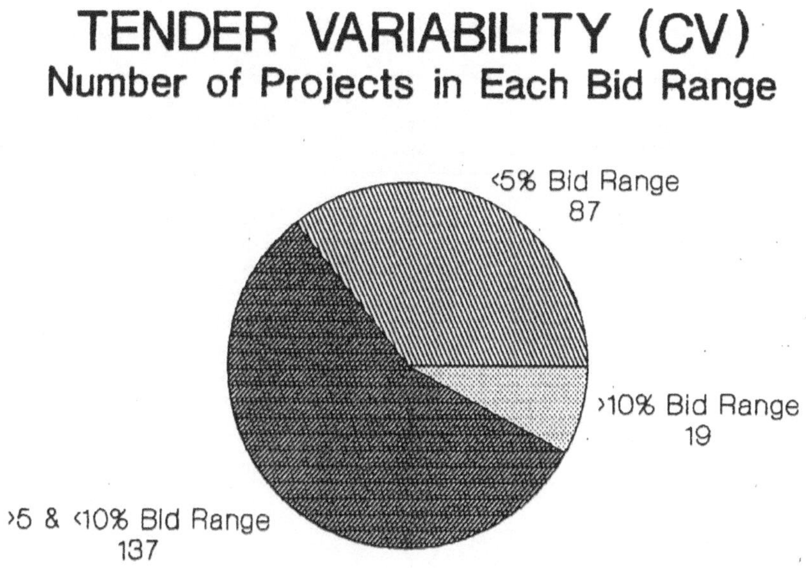
TENDER VARIABILITY (CV)
Number of Projects in Each Bid Range
<5% Bid Range
87
>10% Bid Range
19
>5 & <10% Bid Range
137

Figure 5.20

TABLE 5.9 ANALYSIS ACCORDING TO TENDER RANGE

Tender Range	Number of Projects	Mean Deviation	Mean % Error	CV of Tenders
< 15%	127	11.94	1.25	3.34
>15% & <30%	88	11.29	7.58	6.05
> 30%	28	12.59	5.14	13.02

The majority of the projects (i.e. 52%) were found in the <15% tender range category and in the >5% but <10% CV category (36%). The mean deviation shows a minimal variance across each of the tender ranges, which suggests that the variation of the tenders does not affect the accuracy of the forecast. These results are confirmed by Table 5.10 below.

TABLE 5.10 ANALYSIS ACCORDING TO CV RANGES

CV Range	Number of Projects	Mean Deviation	Mean % Error	Range of Bids
< 5%	87	11.57	5.41	21.77
>5% & <10%	137	11.75	2.71	11.08
> 10%	19	12.97	6.72	55.45

5.3.3.5 The Geographical Location of the Project

The Cape Peninsula was divided into regions labelled A to H. The forecasting accuracy achieved by quantity surveying offices in each of regions was then evaluated. The figures on the following page represent the results, which are summarised in the table below.

TABLE 5.11 ANALYSIS ACCORDING TO THE GEOGRAPHICAL LOCATION

Geographical Location	No. of Projects	Mean Deviation	Mean % Error	CV of Tenders
Area A	86	9.28	3.01	5.96
Area B	3	7.29	4.43	2.97
Area C	5	3.70	-0.56	4.85
Area D	34	20.10	8.66	5.82
Area G	49	13.89	5.59	4.97
Area H	66	10.00	1.99	5.07

The mean deviation varied significantly, from as low as 3.7% (area C) to 20.1% (area D). This wide variance in accuracy suggests that location of the project may influence accuracy. However, the number of projects analysed from area C (and B) was substantially lower than those for the other regions. The small sample size of projects may have resulted in an unreliable result.

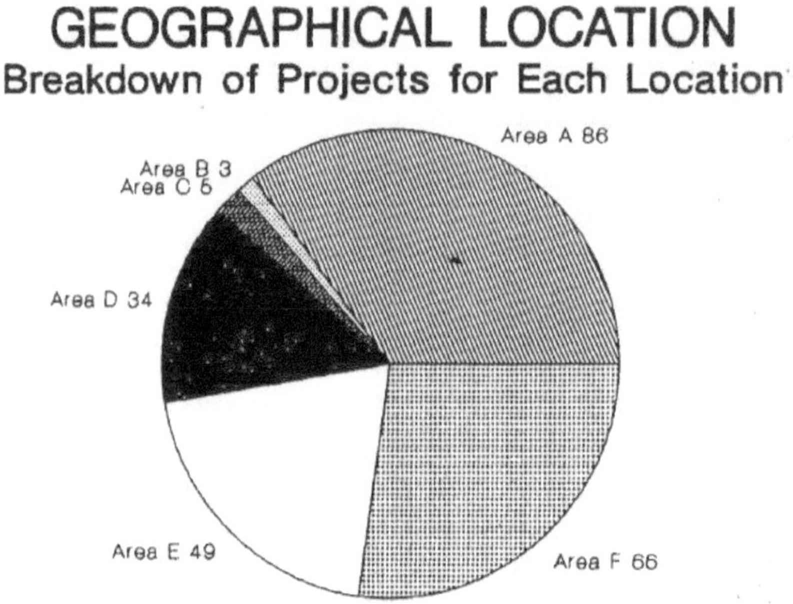

Figure 5.21

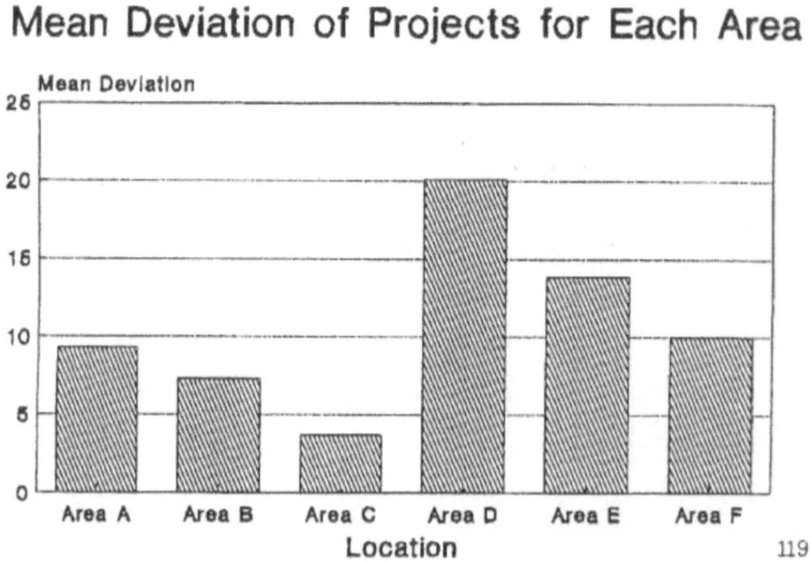

Figure 5.22

GEOGRAPHICAL LOCATION

Mean % Error of Projects for Each Area

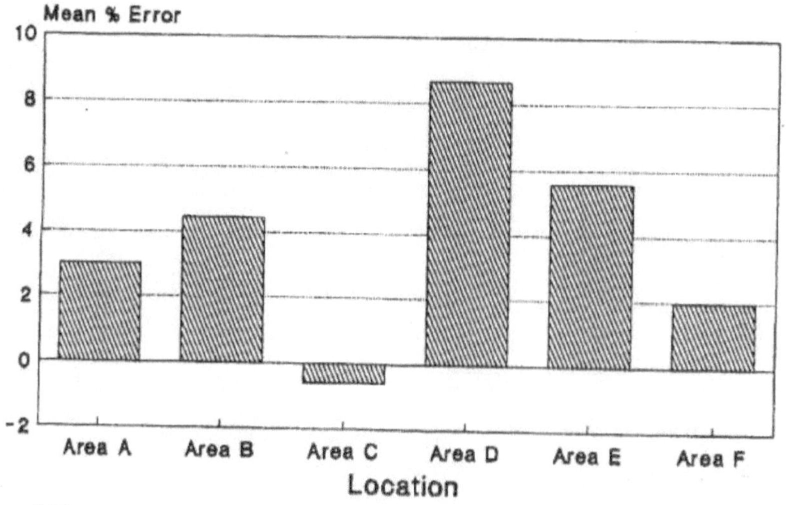

Figure 5.23

CV's of Tenders for Each Location

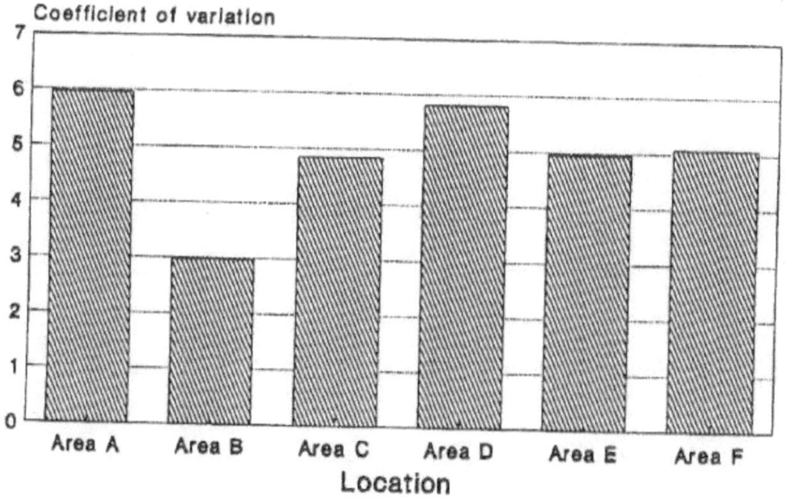

Figure 5.24

Stevens (1983) and Beeston (1975) claim that, in theory, the percentage variability of the bids is the highest possible level of accuracy that the quantity surveyor is able to achieve. Area C's mean deviation is lower than the CV, which indicates that the quantity surveyors were able to estimate the price of the project with a greater degree of accuracy than the contractors.

5.3.3.6 The Percentage of Fixed Sums Allowances

The contracts were classified into groups of those with fixed sum allowances (i.e. provisional and prime cost sums) of less than 10% of the contract value, between 10% and 25%, and greater than 25%.

TABLE 5.12 ANALYSIS ACCORDING TO THE PERCENTAGE OF FIXED SUM ALLOWANCES

% Fixed Sum Allowance	No. of Projects	Mean Deviation	Mean % Error	CV of Tenders
< 10%	67	11.24	2.71	7.56
>10% & <25%	90	7.82	2.01	4.96
> 25%	86	16.34	7.06	4.29

The mean deviation of each category did not indicate any trend and thus the percentage of fixed sum allowances did not appear to affect the accuracy of the forecast.

FIXED SUM ALLOWANCES
Breakdown of Projects in Each Category

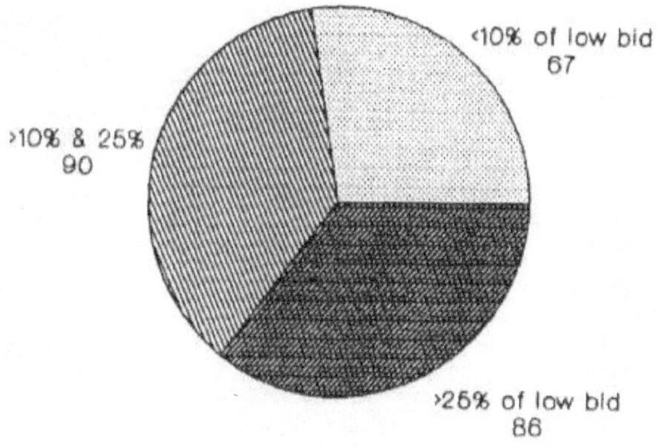

Figure 5.25

Mean Deviation for Each Category

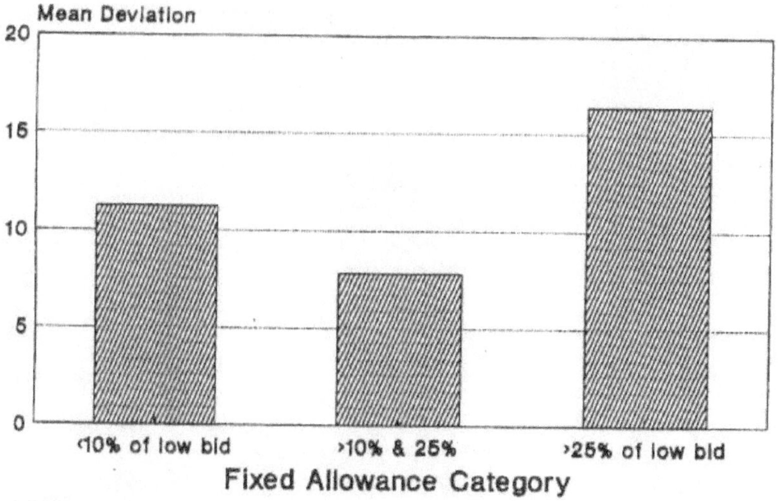

Figure 5.26

FIXED SUM ALLOWANCES

Mean % Error of Each Category

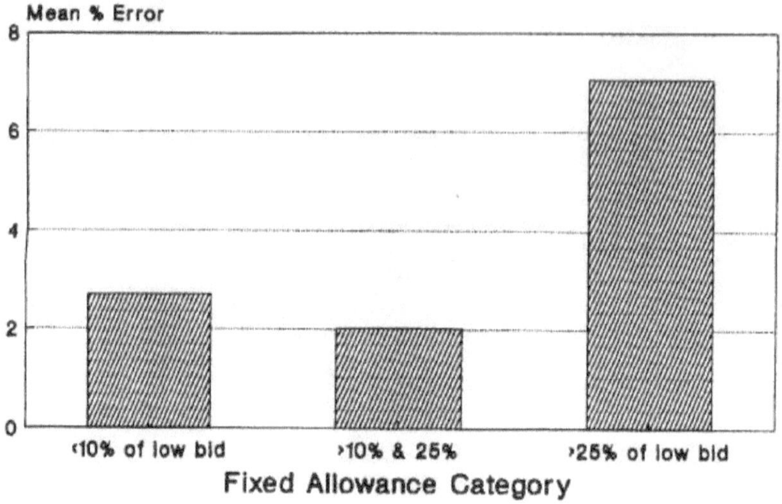

Figure 5.27

CV of Tenders of Each Allowance Category

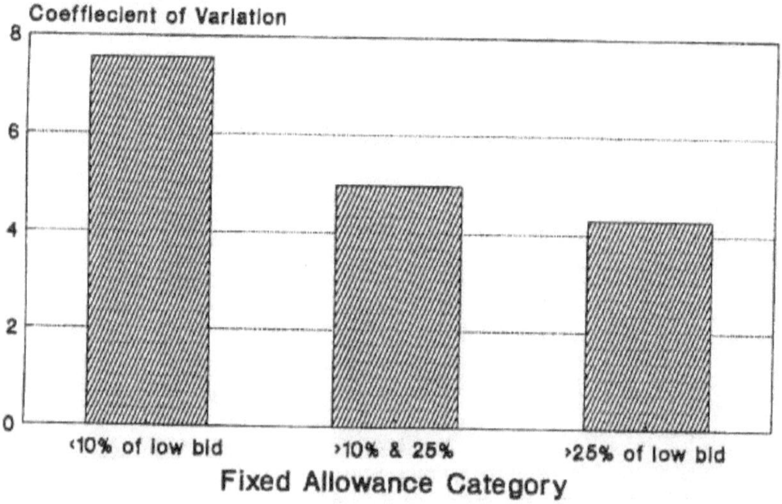

Figure 5.28

5.3.3.7 New or Alterations/Renovations Work

The projects were separated in terms of the nature of the work, i.e. whether the building operations were to an existing structure or not. The consistent level of accuracy, as indicated on the table and figures on the following pages show that the nature of the building operations does not appear to affect the performance of the forecaster or the variability of the tenders.

TABLE 5.13 ANALYSIS ACCORDING TO THE NATURE OF THE WORK

Alteration/ New	Number of Projects	Mean Deviation	Mean % Error	CV of Tenders
Alterations	68	12.47	3.39	5.83
New Work	175	11.51	4.22	5.29

5.3.3.8 The Year of Tender

To determine the effect that the state of the economy has on the quantity surveyor's estimating performance, the projects were separated into their year of tender. The projects are gathered from 1982 to 1991. The graph on the following page illustrates the state (i.e. expansion or contraction) of the economy.

The shaded areas represent the periods of recovery in the economy, which, as explained in chapter three, is characterised by an increase in the work available, a decrease in competition, higher mark ups and more expensive building materials. The unshaded areas depict the periods when the economy is in a recession. It is thought that since the economy affects the contractors tender, it will thus affect the level of accuracy as accuracy is measured in relation to this tender. The mean deviation was then plotted on this graph to determine the effect of the state of the economy.

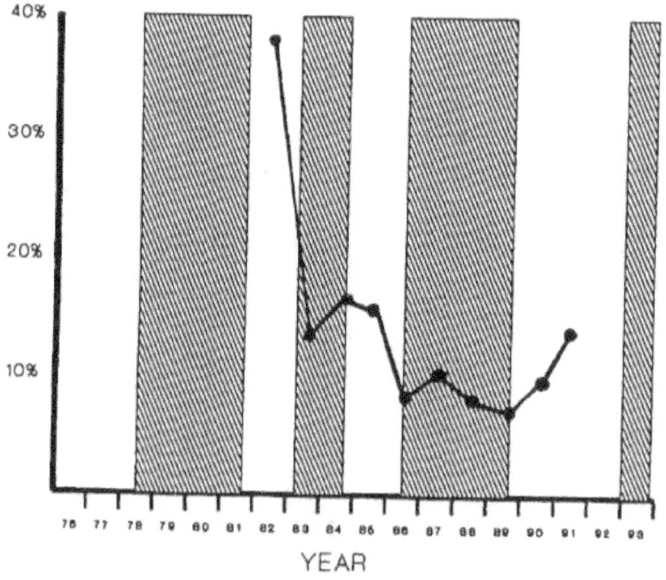

The outcome of the analysis is shown in the table on the following page. The years printed in bold are the shaded years.

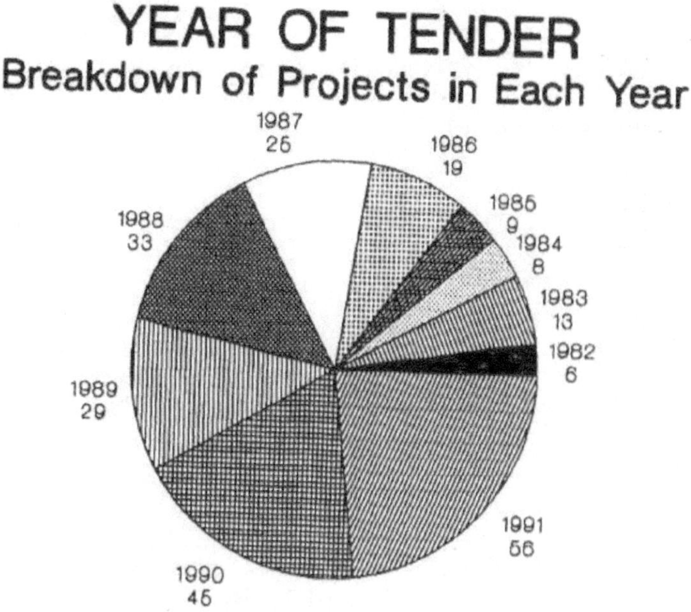

Figure 5.29

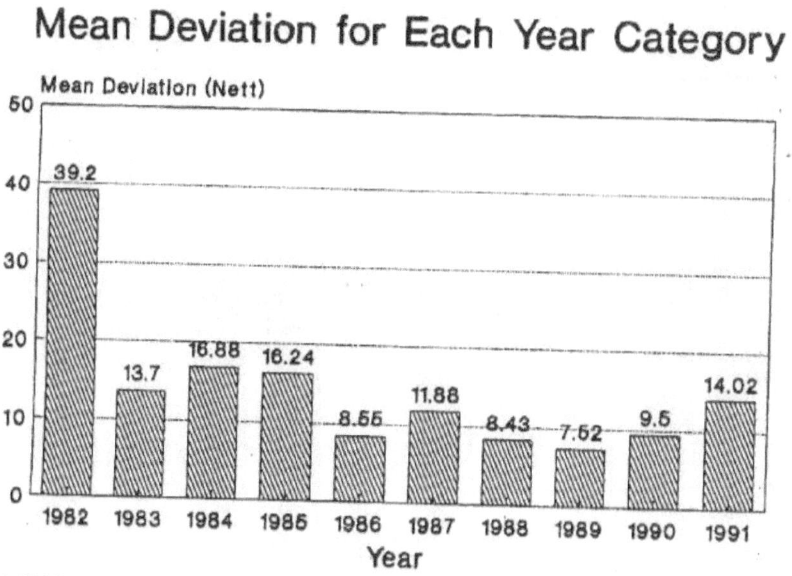

Figure 5.30

YEAR OF TENDER

Mean % Error of Each Year Category

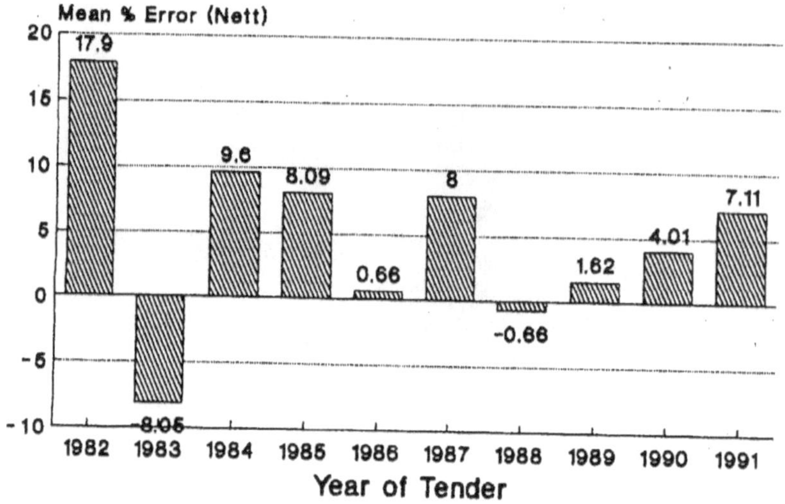

Figure 5.31

CV's of Tender for Each Year Category

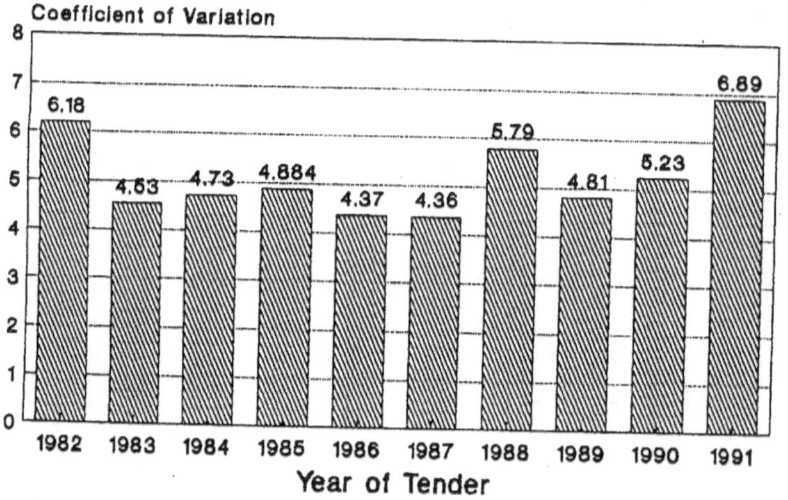

Figure 5.32

TABLE 5.14 ANALYSIS ACCORDING TO THE YEAR OF TENDER

Year of Tender	Number of Projects	Mean Deviation	Mean % Error	CV of Tenders
1982	6	39.20	17.90	6.18
1983	13	13.70	-8.05	4.53
1984	8	16.88	9.60	4.73
1985	9	16.24	8.09	4.88
1986	19	8.55	0.66	4.37
1987	25	11.88	8.00	4.36
1988	33	8.43	-0.66	5.79
1989	29	7.52	1.62	4.81
1990	45	9.50	4.01	5.23
1991	56	14.02	7.11	6.99

The mean percentage error appears to indicate that during the years where the economy is in a state of expansion (1983, 1988) the forecasts are more often underestimated than overestimated or are overestimated almost as often as they underestimated (1986, 1989). The level of accuracy on average tends to be lower where the economy is in a recession. The consistency of the CVs suggests that the state of the economy does not affect the variability of the contractors bids.

5.3.3.9 The Quantity Surveying Office

The expertise of the quantity surveyors could only be measured by the determining the accuracy achieved by each office. The table below summarizes the figures found on the following pages.

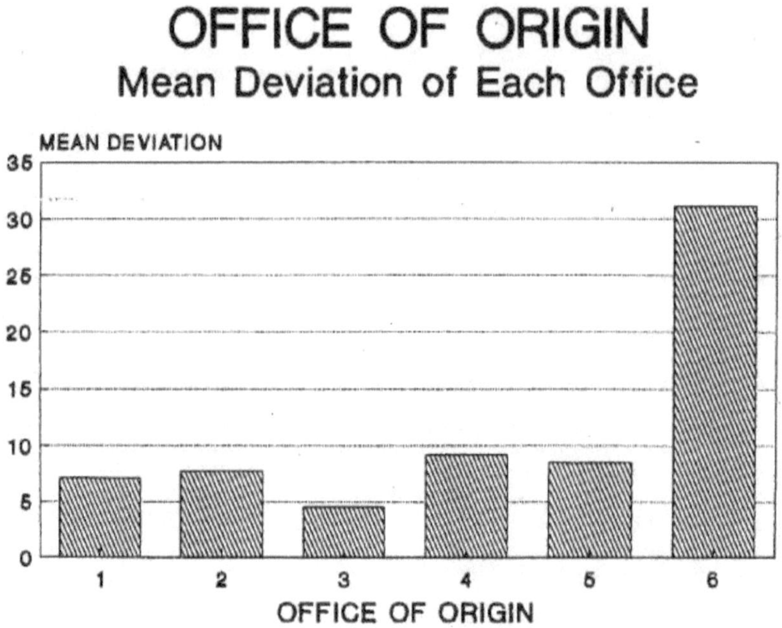
OFFICE OF ORIGIN
Mean Deviation of Each Office
MEAN DEVIATION
35
30
25
20
15
10
5
0
1
2
3
4
5
6
OFFICE OF ORIGIN

Figure 5.34

TABLE 5.15 ANALYSIS ACCORDING TO THE ESTIMATING OFFICE

Estimating Office	Number of Projects	Mean Deviation	Mean % Error	CV of Tenders
Office 1	16	7.11	4.15	4.89
Office 2	14	7.77	1.91	4.95
Office 3	35	4.61	4.13	6.44
Office 4	15	9.23	0.94	5.49
Office 5	24	8.50	-3.13	4.98
Office 6	25	31.16	16.90	5.67

All the offices, except office 5, tend to overestimate on average. Office 3 shows the highest level of accuracy, with a mean deviation of 4.61%, which is substantially higher than the average of 11.78%. Office 6's accuracy (31.16%) is, in contrast, significantly lower than the average. With the exception of offices 3 and 6, the accuracy achieved appears relatively consistent across the remaining offices. This suggests that quantity surveyors in general appear to estimate the tender price with equal proficiency and expertise.

The significantly lower level of accuracy achieved by office 6 affects the average level of accuracy measured of 11.78%. Once this office's data was removed from the sample, the level of accuracy improved from 11.78% to 9.67%.

5.3.3.10 Private and Public Sector Clients

To determine whether the type of client influences the level of accuracy the projects were separated into those measured for the private sector, local government and central government. The table below summarises the results.

TABLE 5.16 ANALYSIS ACCORDING TO TYPE OF CLIENT

Type of Client	Number of Projects	Mean Deviation	Mean % Error	CV of Tenders
Private	179	11.41	3.72	5.15
Local Govt	19	11.57	6.41	5.71
Central Govt	45	13.34	4.04	6.46

The mean deviation of the projects remains relatively constant for each category, suggesting that the type of client has little influence on the accuracy. The variation of the tenders also did not differ significantly, which appears to indicate that the contractors estimate with a relatively constant error, irrespective of the type of client.

5.3.4 Data Estimated by Public (PWD) Quantity Surveying Firms

The public sector data, due to the limited information available, was analysed in terms of the following:

1. The year of tender.
2. The number of bids.

3. The contract value.

4. The type of project.

5. New or alteration/renovations work.

In addition, due to the limited sample size of only 45 projects, the results obtained from the below analyses may not accurately represent the actual performance. The mean deviation was determined to be 14.51%, which was slightly higher than that of the private sector firms, which was found to be 11.78%. This suggests that on average the public sector quantity surveyors are less accurate than those in the private sector. The mean percentage error was also higher, 5.77% as opposed to 4.52%. This indicates that estimates produced by the PWD are more often overestimated than underestimated. The bids obtained for each contract were not supplied, and thus the coefficient of variation of the tenders could not be calculated.

5.3.4.1 The Year of Tender

The PWD projects were tendered in 1990 and 1991. These two years alone are not sufficient to determine whether the year of tender significantly affects the accuracy achieved. The results obtained, however, were as follows :

TABLE 5.17 ANALYSIS ACCORDING TO YEAR OF TENDER

Year of Tender	Number of Projects	Mean Deviation	Mean % Error
1990	19	9.50	4.01
1991	26	14.02	7.11

It can be seen that the mean deviation increased from 1990 to 1991. This increase however, does not necessarily indicate that the market conditions have influenced the accuracy and thus no definite conclusion can be drawn.

5.3.4.2 The Number of Bids

The projects were separated into categories dictated by the number of tenders received. The table below displays the outcome of the analysis of each.

TABLE 5.18 ANALYSIS ACCORDING TO THE NUMBER OF BIDS

Number of Bids	Number of Projects	Mean Deviation	Mean % Error
< 3	6	25.65	7.87
< 5	10	20.11	8.27
>5 & <10	33	13.29	5.16
> 10	6	10.10	7.07

The above shows the mean deviation to increase as the number of bids decreases, suggesting a positive correlation with accuracy. However due to the small, and thus unreliable, sample sizes, no definite conclusion can be drawn.

NUMBER OF BIDS
Number of Projects in Each Bid Category

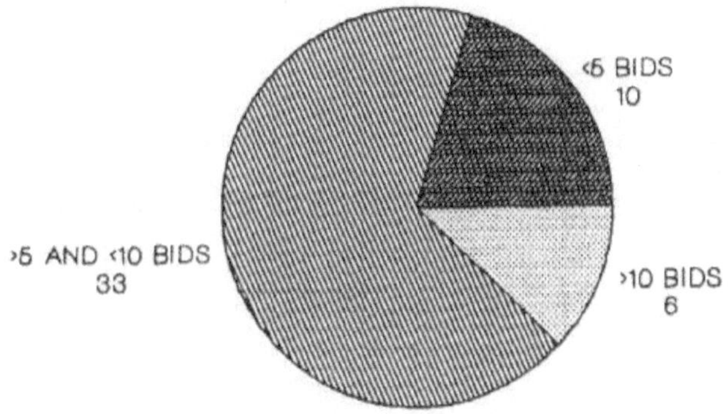

Figure 5.35

Mean Deviation for Each Bid Category

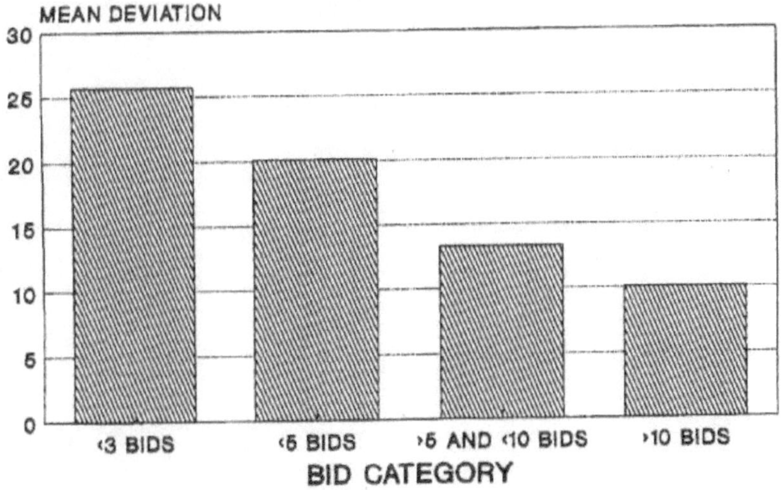

Figure 5.36

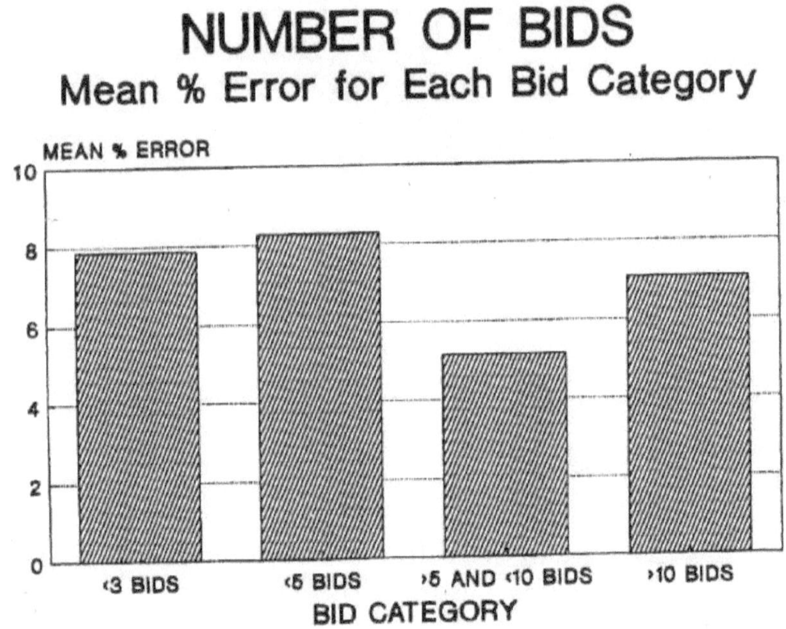
NUMBER OF BIDS
Mean % Error for Each Bid Category
MEAN % ERROR
10
8
6
4
2
0
<3 BIDS
<5 BIDS
>5 AND <10 BIDS
>10 BIDS
BID CATEGORY

Figure 5.37

CONTRACT VALUES
Number of Projects in Each Size Category

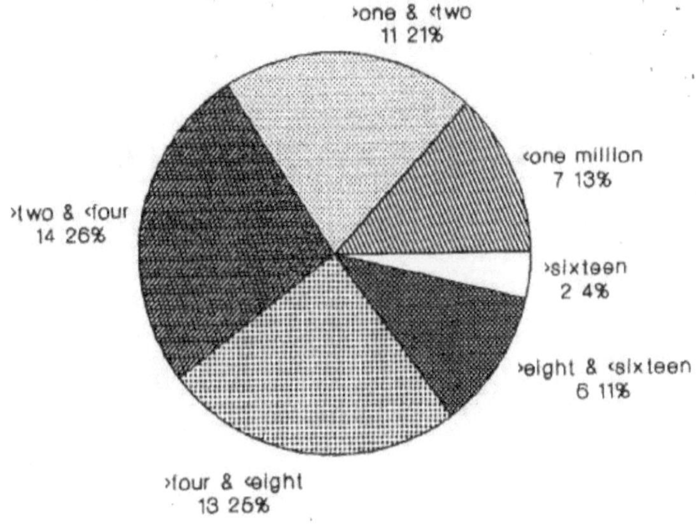

Figure 5.38

Mean Deviation for Each Value Category

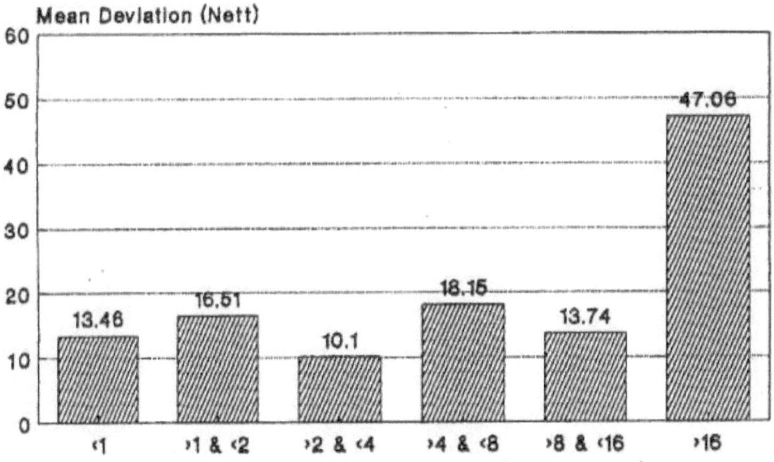

Figure 5.39

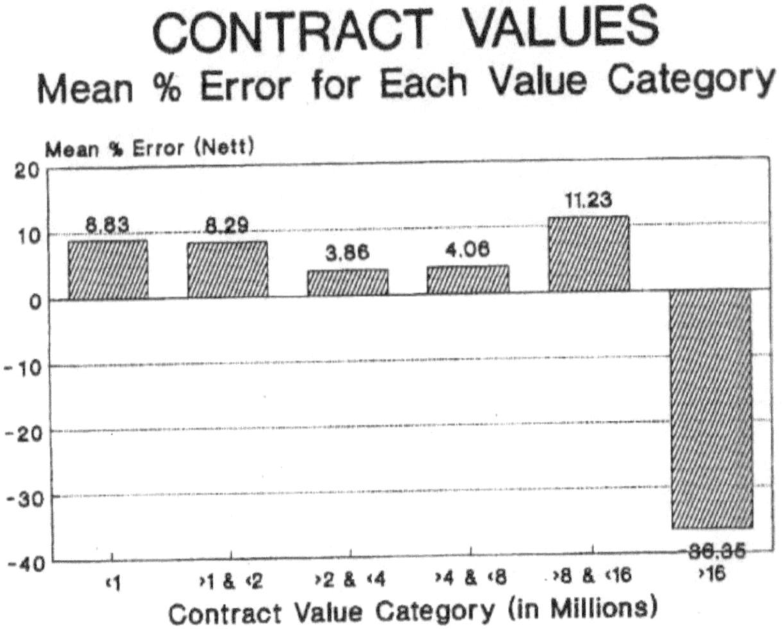
CONTRACT VALUES
Mean % Error for Each Value Category
Mean % Error (Nett)
20
10
0
-10
-20
-30
-40
8.83
8.29
3.86
4.06
11.23
36.35
‹1
›1 & ‹2
›2 & ‹4
›4 & ‹8
›8 & ‹16
›16
Contract Value Category (in Millions)

Figure 5.40

5.3.4.3 The Contract Value

The projects were grouped into four value categories, the fourth category was then analysed in three groups.

TABLE 5.19 ANALYSIS ACCORDING TO THE CONTRACT VALUE

Project value (R millions)	No. of Projects	Mean Deviation	Mean % Error
< 1	7	13.46	8.83
>1 & <2	11	16.51	8.29
>2 & <4	14	10.10	3.86
> 4	13	18.15	4.06
>4 & <8	6	13.74	11.23
>8 & <16	2	47.06	-36.35
> 16	5	11.87	11.63

The fluctuating mean deviation shows no apparent trend to exist. At the two to four million rand contract value category the forecasts are most accurate.

5.3.4.4 The Type of Contract

The contracts supplied by the PWD consisted of only two types of contracts, i.e. commercial and residential. In addition to this, the fact that the majority (85%) of the projects were commercial developments it precluded any reliable deductions from being drawn.

TABLE 5.20 ANALYSIS ACCORDING TO THE TYPE OF PROJECT

Type of Project	No. of Projects	Mean Deviation	Mean % Error
Commercial	34	13.61	9.87
Residential	6	23.22	-7.65

The large discrepancy between the two work groups suggests that the type of project may affect the level of accuracy.

5.3.4.5 The Nature of the Work

The new work and alterations or renovations were analysed to determine if the nature of the work had any impact on the accuracy achieved.

TABLE 5.21 ANALYSIS ACCORDING TO THE NATURE OF THE WORK

Nature of the Work	No. of Projects	Mean Deviation	Mean % Error
New Work	33	12.15	4.34
Alterations	12	21.01	9.71

The level of accuracy increases significantly for alterations projects, suggesting that these types of projects negatively affect the accuracy achieved. The mean percentage error indicates that the both types of projects are on average more frequently overestimated.

5.4 Comparison of PWD and Private Sector Data

Due to the limited nature of the PWD data, all the factors analysed for the private sector projects could not be performed on the PWD data. It also should be noted that due to the relatively small sample of the PWD data (45 projects) the analyses performed on these projects may be unreliable. The following is a comparison of those factors which were analysed for both sets of data.

5.4.1 Level of Accuracy Achieved

The private sector estimated the contract price with an average accuracy of 11.78% (mean net deviation), while the PWD averaged 14.51%. The mean percentage error for each sector of estimators was 4.52% for the private sector and 5.77% for the PWD. The fact that mean percentage errors for both sectors was positive indicates that the quantity surveyors tend to overestimate more often than underestimate the price of the contract.

5.4.2 The Value of the Contract

From both groups of data it was apparent that no distinctive trend exists relating the value of the contract to the level of accuracy achieved.

5.4.3 The Type of Project

The PWD data contained only two different types of projects, (commercial and residential) which precluded a more reliable analysis of the effect of project type. The mean deviation of these two types, however, varied quite significantly. The accuracy of the commercial developments was measured at 13.61%, while the residential was 23.22%, indicating that the project type may influence accuracy. The private sector data showed two types to be slightly more inaccurate than the others. These types were the educational and religious facilities contracts. The residential projects showed a mean deviation of 10.19% which is substantially lower than that obtained by the PWD.

5.4.4 The Number of Bids

The private sector projects indicated no trend to exist, the level of accuracy remained consistent for all categories analysed. From the PWD data it appeared that the accuracy improves as the number of bidders increases.

5.4.5 The Nature of the Work

The mean deviation of the PWD projects showed a significant difference in the accuracy - 12.15% for new work and 21.01% for alterations. This appears to suggest that the nature of the work influences the accuracy. The private sector data did not confirm this, showing no trend to exist.

5.4.6 The Year of the Tender

During the periods of slower economic growth the private sector quantity surveyors tended to overestimate the tender price. The PWD data obtained was only from two years, and thus the effect of the prevailing market conditions could not be ascertained.

5.5 Comparison of Actual and Perceived

The following compares the perceived levels of accuracy and the influencing factors observed in the questionnaire with the actual results found in the empirical study. Since the questionnaires were only distributed to private quantity surveying firms, the empirical study is only to be compared with the private sector data.

5.5.1 Levels of Accuracy

The respondents to the questionnaire perceive their accuracy to improve from 12.24% (inception) to 6.33% (tender stage) over the design process. The empirical study showed the accuracy at tender stage to be 11.78%, which is significantly higher than the 6.33% envisaged. This substantiates the hypothesis which proposes that quantity surveyors are unaware of the magnitude of their forecasting error.

5.5.2 The Type of Project

The type of project is considered by 58.6% of the respondents to influence the accuracy achieved. The effect of the project type is believed to result from the experience with such projects. Commercial projects are associated with the highest level achieved. The empirical study showed some types of projects to exhibit a higher level of error than others, which appeared to suggest that the type of project may affect the level of accuracy achieved. The accuracy of commercial developments, which was thought to be the type of project on which accuracy was highest, was measured at 13.95%. This is below the average level of accuracy (11.78%) for all projects, indicating that quantity surveyors are not aware of their actual performance.

5.5.3 The Location of the Project

The majority (63%) of the respondents considered the project location to exert "little" or "some influence" over accuracy. The empirical study, however, showed the accuracy to vary significantly with the location, from a high as 3.7% to as low as 20.1%. These results seem to suggest that the location of the project does affect the level of accuracy.

5.5.4 The Value of the Project

Approximately 77% of the respondents considered the value of the project to have "little" or "some" influence over the accuracy achieved. The results of the analysis of the data confirmed these assumptions. No trend could be established from the levels of accuracy of the value categories.

5.5.5 The State of the Market

The prevailing economic conditions are considered by almost 77% of the surveyors to have a significant or critical effect on the level of accuracy. The data show no definitive trends, although it appeared that during the periods of expansion of the economy, the accuracy tended to be better. In addition the mean percentage error indicated that during such times, forecasts tended to be more often, or as frequently underestimated as overestimated.

5.5.6 The Quantity Surveying Office

The only means of determining the effect of the degree of expertise on accuracy was to measure the levels of accuracy achieved by each quantity surveying office. The opinion survey indicated that the majority (87%) considered expertise to exert "significant" or critical influence over the accuracy achieved. The empirical study showed the accuracy of two of the offices to deviate widely from that of the remaining firms, all of whose accuracy was relatively consistent. Office 3 showed a significantly higher level of accuracy (4.61% mean deviation) than the average (11.78%), while office 6 had a substantially lower level of accuracy (31.16%). These results appeared to confirm the opinions of the respondents.

5.6 Conclusions

The results of the comparison of the actual and perceived levels of accuracy support Bennett et al.'s (1981) and Ogunlana's (1991) premise that quantity surveyors are unaware of the magnitude of their errors. In addition those factors considered to exert the greatest influence over the degree of error were identified. These factors are thought to be :

1. project type,
2. geographical location,
3. quantity surveying office, and possibly
4. the nature of the work and
5. the year of tender (i.e. market conditions).

The following reasons are suggested for the possible influence the above factors exert over accuracy :

1. The Nature of the Project

   Some types of projects are more repetitive in nature than others, which could result in the estimating task being more analytical and predictable, and thus the price may be easier to assess. In addition, those offices which specialize or have significantly greater experience with a particular type of project may find that their accuracy is noticeably better with this type. This is likely to due to a larger and more current cost data base which will have been accumulated, as well as the experience gained with such types.

2. Geographical Location

The location of the project can be affected by the particular Local Authority's regulations. For example a project to be carried out in the city centre will have different price implications to one in an industrial area located relatively far from the labour and material supply. The ability of the quantity surveyor to anticipate the effect that the location will have on the contractor's price will influence the accuracy of his/her forecast.

3. The Quantity Surveying Office

Two of the offices showed significantly higher and lower levels of accuracy. The accuracy of the office as a whole could be affected by the any number of factors including the expertise and experience of the individual forecasters, the amount and quality of cost data stored in the office or the estimating techniques used.

4. The Nature of the Work

The PWD projects showed a marked difference between the new and alterations work. The alterations projects were significantly overestimated. This may be due to the uncertain nature of the existing and surrounding structures and building operations which are to follow.

5. The Year of Tender

The prevailing state of the economy largely influences the mark up chosen by the contractor, as well as his labour and material costs. The estimator must anticipate the degree to which the present market conditions will affect the contractor's bid.

It should be noted however, that no statistical testing was carried out on this data to verify the apparent trends, and thus any conclusions drawn are based only on visual observations.

CHAPTER SIX : SUMMARY AND CONCLUSIONS

CHAPTER SIX

SUMMARY AND CONCLUSIONS

This dissertation sets out to establish the most influential factors affecting the accuracy of the price forecast. Since no studies have been conducted in South Africa to determine what the actual levels achieved are, it is presumed that quantity surveyors are largely unaware of the degree of accuracy that they attain. To validate this problem, an opinion survey was conducted to ascertain the perceived levels of accuracy. These levels were then compared with the actual levels determined by an empirical analysis of tenders and forecasts.

A discussion of the theory of accuracy and error is contained chapter two. This chapter considers the importance of accuracy in the cost control process, as well as its measurement and the various forecasting techniques available.

Chapter three investigates the supposition that estimating accuracy is influenced by certain variables or factors. These factors, their effects and possible relationships to estimating accuracy are described. It was found that the majority of the existing research to date has failed to prove a uniform and statistically significant relationship between these factors and accuracy. From this research, the following factors (in no particular order) were found to exert some influence over the accuracy of the forecast:

1. the project type,
2. the project size,
3. the geographical location,
4. the market conditions,
5. the number of bidders,
6. the historical cost data available and
7. the expertise of the forecaster.

A study of existing literature was conducted in chapter four, reviewing much of the published research in the field of price forecasting accuracy.

The final chapter sets out to validate the hypothesis that the most influential factors which exert an influence over the estimating accuracy can be identified. An opinion survey was initially carried out to determine the general perceptions of accuracy and its associated factors. It was found that accuracy is perceived to increase from 12.24% at inception to 6.33% at tender. The factors considered by the respondents to exert a significant influence over accuracy were the amount of design information available, the prevailing market conditions and the expertise of the forecaster.

An empirical study of construction projects carried out over the last ten years was conducted to determine the validity of the perceptions of accuracy and the factors which influence it.

The lowest tender was compared with the quantity surveyor's forecast to establish the actual levels of accuracy achieved. The data was then analysed to ascertain which factors affected the accuracy of these bids. From the factors analysed, the following were found to exert the greatest influence over the level of accuracy :

1. the project type,
2. the geographical location,
3. the quantity surveying office, and possibly
4. the nature of the work (i.e. new or renovations) and
5. the market conditions.

Most of the above factors (project size, market conditions, geographical location, and possibly the quantity surveying office if considered as a measure of expertise) correspond with those from existing research.

In addition, when the actual and perceived average levels of accuracy were compared, the discrepancy between the two figures indicates that the quantity surveyors believe their accuracy to be significantly higher than it in actual fact is. This shows that in general quantity surveyors possess an overly optimistic impression of their estimating performance.

The ignorance of the actual levels of accuracy achieved may cause such errors in the forecast to inadvertently persist. An awareness of the forecasting accuracy attained and the factors which influence it may contribute towards enhanced estimating performance. As a means of ensuring that the accuracy is correctly assessed a system of monitoring the mean deviation of the forecasts on a regular basis is recommended. This would provide constant feedback on the true levels of accuracy attained. Such feedback may also create an awareness of the factors that exert an influence over the estimate's accuracy. The quantity surveyor would thus be more informed and more capable of identifying the potential problem areas within a project, for which he/she could make sufficient allowances. This would then lead to an improved level of accuracy.

An improvement in the levels of accuracy of price forecasts would enable the quantity surveyor to be more effective in ensuring that the cost control process is able to fulfill all of its objectives.

LIST OF REFERENCES

LIST OF REFERENCES

Allman, I. (1988) Significant items estimating. *Chartered Quantity Surveyor*, September, pp. 24-25.

Ashworth, A. (1988) *Cost Studies of Buildings*. Longman Scientific and Technical.

Ashworth, A., Neale, R.H., and Trimble, E.G. (1980) An Analysis of the accuracy of some builder's estimating. *The Quantity Surveyor*, April, pp. 65-70.

Ashworth, A. and Skitmore, R.M. (1983) Accuracy in estimating. *Occasional Paper No. 27*, Chartered Institute of Building, London.

Beeston, D.T. (1975) One statistician's view of estimating. *Building Technology and Management*, March, pp. 33-37.

Beeston, D.T. (1987) A future for cost modelling, in *Building Cost Modelling and Computers* (Ed. P.S. Brandon), E. and F.N. Spon Ltd., London, pp. 17-24.

Bennett, J., Morrison, N.A.D. and Stevens, S.D. (1981) Cost planning and computers. *Property Services Agency* Library, London.

Bennett, J. and Ormerod R.N. (1984) Simulation applied to construction projects. *Construction Management and Economics*, Vol. 2 No. 3, pp. 225-263.

Bennett, J. (1984) Cost data and the QS, *Chartered Quantity Surveyor*, April, pp. 345-346.

Bennett, J. and Barnes, M. (1979) Outline of a theory of measurement, *Chartered Quantity Surveyor*, Vol. 2, October, pp. 53-56.

Betts, M. and Gunner, J. (1989) The accuracy of a consultant quantity surveyor's building price prediction. *Occasional Paper No. 3/89*, School of Building and Estate Management, National University of Singapore.

Birnie, J. and Yates, A. (1991) Cost prediction using decision/risk analysis methodologies, *Construction Management and Economics*, Vol. 9, pp. 171-186.

Bowen, P. A. and Edwards, P.J. (1985) Cost modelling and price forecasting: from a deterministic to a probabilistic approach. Occasional Paper No. 5, *Occasional Paper Series*, Department of Quantity Surveying and Building Economics, University of Natal, Durban.

Bowen, P.A. and Edwards, P.J. (1985) Cost modelling and price forecasting: practice and theory in perspective, *Construction Management and Economics*, Vol. 3, pp. 199-215.

Bowen, P.A., Wolvaardt, J.S. and Taylor, R.G. Cost modelling : a process modelling approach, in *Building COst Modelling and Computers* (Ed. P.S. Brandon), E. and F.N. Spon Ltd, London, pp. 387-396.

Bowen, P.A. (1992) A communication based approach to price modelling and the price forecasting in the design phase of traditional building procurement process in South Africa. Unpublished Ph.D. Thesis, University of Port Elizabeth Forthcoming.

Brandon, P. and Newton, S. (1986) Improving the forecast, *Chartered Quantity Surveyor*, May, pp. 24-26.

Cartlidge, D.P. and Mertens, I.N. (1982) *Practical cost planning*. Hutchinson and Co., London, pp.9-16.

Department of the Environment (Directorate of Quantity Surveying Services) (1980) *Construction Cost Data Base*. Second Annual Report. Property Services Agency Library, London.

Drew, D.S., and Skitmore, R.M. (1990) Analysing bidding

performance; measuring the influence of contract size and type, in Proceedings of the C.I.B. W-55/65 International Symposium on *Building Economics and Construction Management*, Sydney, March, Vol. 6, pp. 129-139.

Drew, D.S. and Skitmore, R.M. (1992) Competitiveness in bidding : a consultant's perspective, *Construction Management and Economics*, Vol. 10 No. 3, May, pp. 227-247.

Ferry, D.J. and Brandon, P.S. (1991) *Cost Planning of Buildings*. Sixth Edition, Blackwell Scientific Publications, London.

Flanagan, R. and Norman, G. (1978) The relationship between construction price and height. *Chartered Surveyor*, Vol. 4, pp. 69-71.

Flanagan, R. and Norman, G. (1982a) Risk analysis - an extension of price prediction techniques for building work. Management, University of Reading, Reading. *Construction Papers*, Vol 1, No3, pp. 27-43.

Flanagan, R. and Norman, G. (1982b) Making good use of low bids. *Chartered Quantity Surveyor*, March, pp. 226-227.

Flanagan, R. and Norman, G. (1983) The accuracy and monitoring of quantity surveyor's price forecasting for building work. *Construction Management and Economics,* Vol 1, pp. 157-180.

Flanagan, R. and Stevens, S. (1990) Risk analysis, in *Quantity Surveying Techniques : New Directions* (Ed. P. S. Brandon), Blackwell Scientific Publications, London.

Gilmour, J. and Skitmore, R.M. (1989) A new approach to early stage estimating, *Chartered Quantity Surveyor,* May, pp. 36-38.

Gonzalez, A.E., Smith, C. and Rincon, B. (1988) Cost estimating at Corpoven. Transactions, *American Association of Cost Engineers,* pp. B.8.1-B.8.4.

Gunner, J. and Betts, M. (1990) Price forecasting performance by design team consultants in the Pacific rim, in Proceedings of the C.I.B. W-55/65 International Symposium on *Building Economics and Construction Management,* Sydney, March, Vol. 3, pp. 302-312.

Harmer, S. (1983) Identifying significant BQ items, *Chartered Quantity Surveyor,* October, pp. 95-96.

Hemphill, R.B. (1968) A method for predicting the accuracy of a construction cost estimate. Transactions, *12th National*

Meeting of the American Association of Cost Engineers, Houston, June, pp. 20.1-20.18.

Huxley, A. L. (1991) Building design estimating accuracy: what's reasonable? Transactions, *American Association of Cost Engineers*, pp. M.1.1-m.1.5.

Lau, A. H. and Lau, H. (1986) A system for improving the accuracy of cost estimates. *Production and Inventory Management*, Fourth Quarter, pp. 89-100.

Makridikas, S., Wheelwright, S. C. and McGee, V. E. (1983) *Forecasting : Methods and Applications*, John Wiley and Sons, Canada.

Mathur, K. S. (1982) A probabilistic planning model, in *Building Cost Techniques : New Directions* (Ed. P.S. Brandon), E. & F. N. Spon Ltd., London, pp. 181-191.

McCaffer, R., McCaffery, M. J. and Thorpe, A. (1984) Predicting the tender price of buildings during early design : method and validation. *Journal of the Operational Research Society*, Vol. 35, No. 5, pp. 415-424.

Morrison, N. and Stevens, S. (1980) A construction cost data base. *Chartered Quantity Surveyor*, June, pp. 313-315.

Morrison, N. (1983) *The Cost Planning and Estimating Techniques*

Employed by the Quantity Surveying Profession. Unpublished PhD Thesis, Department of Construction Management, University of Reading, Reading.

Morrison, N. (1984) The accuracy of quantity surveyor's cost estimating. *Construction Management Economics*, Spring, pp.57-75.

Newton, S. (1988) Cost modelling techniques in perspective. Transactiona, *American Association of Cost Engineers*, pp. B.7.1-B.7.7.

Newton, S. (1984) *Expert systems and the quantity surveyor.* Surveyors Publications, London.

O'Dean, D. W. (1984) Estimating for building projects. *The Building Economist, June, pp. 11-29.*

Ogunlana, S. O. (1989) *Accuracy in design cost estimating.* PhD Thesis, Longhborough University of Technology.

Ogunlana, S. O. (1991) Learning from experience in design cost estimating. *Construction Management and Economics*, 9, pp. 133-150.

Ogunlana, S. O. and Thorpe, A. (1987) Design phase cost

estimating : the state of the art. *International Journal of Construction Management and Technology*, 2, 4, pp. 34-47.

Ogunlana, S. O. and Thorpe, A. (1990) A new direction for design cost estimating, in Proceedings of the C.I.B. W-55/65 International Symposium on *Building Economics and Construction Management*, Sydney, March, Vol. 2, pp. 206-216.

Ogunlana, S. O. and Thorpe, A. (1991) The nature of estimating accuracy : developing correct assumptions. *Building and Environment*, Vol. 26, No. 2, pp. 77-86.

Raftery, J. (1984) *An Investigation into the Suitability of Cost Models in Building Design*, PhD Thesis, Department of Surveying, Liverpool Polytechnic.

Raftery, J. (1987) The state of cost/price modelling in the UK construction industry: a multicriteria approach, in *Building Cost Modelling and Computers* (Ed. P.S. Brandon), E. & F. N. Spon Ltd., London, pp. 49-71.

Ray-Jones, A. and Clegg, D. (1976) *Construction Indexing Manual* RIBA, London.

Runeson, G. (1988) An analysis of the accuracy of estimating and the distribution of tenders. *Construction Economics and Management*, 6, pp. 357-370.

Runeson, G. and Bennett, J. (1983) Tendering and the price level in the New Zealand building industry. *Construction Papers* Vol. 2, No. 2, pp. 29-35.

Seeley, I. H. (1983) *Building Economics.* Macmillan, London.

Skitmore, R. M. (1981) Why do tenders vary? *Chartered Quantity Surveyor*, December, pp. 128-129.

Skitmore, R. M. (1981) *Bidding Dispersion - An Investigation into a method of measuring the accuracy of Building Cost Estimates.* Unpublished MSc Thesis, Department of Civil Engineering, University of Salford.

Skitmore, R. M. (1986) *Towards an Expert Building Price Forecasting System.* Surveyors Publications, London.

Skitmore, R. M. (1987a) The effect of project information on the accuracy of building price forecasts, in *Building Cost Modelling and Computers* (Ed. P.S. Brandon), E. & F. N. Spon Ltd., London, pp. 327-336.

Skitmore, R. M. (1987b) The distribution of construction project bids, in Proceedings of the C.I.B. Fourth International Symposium on *Building Economics*, Copenhagen, Session D, pp. 171-183.

Skitmore, R. M. and Tan, S. H. (1987) Factors affecting accuracy of engineers estimates. Research report, Department of Civil Engineering, University of Salford.

Skitmore, R. M. (1988) Factors affecting accuracy of engineers' estimates. Transactions, *American Association of Cost Engineers*. Vol. 30, No. 12, December, pp. 12-17.

Skitmore R. M. and Patchell, B. (1990) Developments in contract price forecasting and bidding techniques, in *Quantity Surveying Techniques: New Directions* (Ed. P.S. Brandon), Blackwell Scientific Publications, London.

Skitmore, R. M., Stradling, S., Tuohy, A. and Mkwezalamba, H. (1990) *The Accuracy of Construction Price Forecasts*, The University of Salford.

Stevens, S. D. (1983) *The Use of Price Data from Analysis of Bills of Quantities in Construction Price Estimating.* Unpublished PhD Thesis, Department of Construction Management, University of Reading, Reading.

Stevens, G. and Davis, T. (1988) How accuract are capital cost estimates? Transaction *American Association of Cost Engineers*, pp. B.4.1-B.4.5.

Tan, S. H. (1988) *An Investigation into the Accuracy of Cost Estimates during the Design Stages of Construction Projects*, Unpublished BSc Thesis, Department of Civil Engineering, University of Salford.

True, N. F. (1988) Determining the accuracy of a cost estimate. Transactions, *American Association of Cost Engineers*, pp. T.2.1-T.2.10.

Wilson, O. D., Sharpe, K. and Kenley, R. (1987) Estimates given and tenders received : a comparison. *Construction Management and Economics*, 5, pp. 211-226.

Wilson, O. D. and Sharpe K. (1988) Tenders and estimates : a probabilistic model. *Construction Management and Economics*, 6 pp. 225-245.

Zahry, M. (1988) *Capital Cost Predictions by Multi-variate Analysis*, Unpublished MSc Thesis, Department of Architecture and Building Science, University of Strathclyde, Glasgow.

APPENDICES

UNIVERSITY OF CAPE TOWN

Department of Construction Economics and Management

University of Cape Town · Private Bag Rondebosch 7700
Centlivres Building · Telephone: 650-3443
Telefax No. (021) 650-3726
Head – Professor A.J. Stevens

10 June 1992

Dear Sir

QUESTIONNAIRE ON ACCURACY IN DESIGN COST ESTIMATING

Please assist Ms Donald by participating in a questionnaire survey in respect of the accuracy of cost advice provided by quantity surveyors to clients and architects.

The purpose of this survey is to gather information for use in her final year research project for the Degree of Bachelor of Science in Quantity Surveying at the University of Cape Town.

The information gleaned from respondents will be processed in aggregate form only, and confidentiality is ensured. The results of this study will be made available to participants.

Thank you for your support, without which this project would not be possible.

Yours sincerely

Signed by candidate

PAUL BOWEN
Professor

The University of Cape Town rejects racism and racial segregation and strives to maintain a strong tradition of non-discrimination with regard to race, religion and gender in the constitution of its student body, in the selection and promotion of its staff and in its administration.

University of Cape Town
Private Bag
Rondebosch
7700
Centlivres Building

10 June 1992

J.J. Schneid and Partners
P.O. Box 6941
Roggebaai
8012

Dear Sir

QUESTIONNAIRE ON ACCURACY IN DESIGN COST ESTIMATING

Enclosed herewith you will find a questionnaire relating to the accuracy of pre-tender cost estimates and the factors which affect it. I would be extremely grateful if you could answer by completing this form.

The purpose of this survey is to gather information for use in my research thesis towards requirements for the degree of Bachelor of Science in Quantity Surveying. The thesis is entitled "An Investigation into the Accuracy of Design Cost Estimates".

The confidentiality of information provided will be strictly maintained.

Thanking you in anticipation for your support.

Yours sincerely

Signed by candidate

G. DONALD

QUESTIONNAIRE ON ACCURACY IN DESIGN COST ESTIMATING

FROM : GAIL DONALD

(BSc (QS) UNDERGRADUATE STUDENT

AT THE UNIVERSITY OF CAPE TOWN)

RESPONDENT : ....

DATE : ....

ACCURACY OF DESIGN COST ESTIMATING

This research is designed to find out the opinions of practising design cost estimators (Quantity Surveyors) about certain issues that affect the accuracy of cost estimates. Response to the questions will be treated as confidential.

1. Which estimating technique is normally used by your office at the following design stages?

   Inception ---

   Feasibility ---

   Sketch Design ---

   Detail Design ---

   Tender --

2. What type of cost data (for example: in-house data, published price books, specialist quotes, etc.) is most frequently used during the following design stages in conjunction with the estimating techniques stated in Question 1?

   Inception ---

   Feasibility ---

   Sketch Design ---

   Detail Design ---

   Tender --

3. In what form does your office store "in-house" cost data?

(For example: cost analyses, updated bills, etc.)

4. What are the actual levels of accuracy of your forecasts, using the techniques stated in question 1, at each of the following design stages?

Inception ___________________%

Feasibility _________________%

Sketch Design _______________%

Detail Design _______________%

Tender ______________________%

5. Do you feel that the accuracy of your forecasts improves significantly once detailed design information becomes available?

```
+--+              +--+
|  | YES          |  | NO
+--+              +--+
```

Why?____

____

6. Please state what design information is provided by the architect at the following design stages:

Inception ____

Feasibility ____

Sketch Design _______________________________________

Detail Design _______________________________________

Tender ____

7. Do you feel that there is sufficient design information at the Inception Stage to perform a "realistic" forecast?

```
+--+           +--+
|  | YES       |  |  NO
+--+           +--+
```

If no, at what stage do you believe that there is sufficient design information to perform a "realistic" forecast: ______________________________

8. How often do you compare your estimated design costs with the accepted tender figure?

```
            +--+                      +--+
    Never   |  |    Occasionally      |  |
            +--+                      +--+
            +--+                      +--+
Frequently  |  |          Always      |  |
            +--+                      +--+
```

9. Do you feel that your level of accuracy is significantly higher on certain types of projects (for example housing, schools, civil engineering works, etc)?

```
+--+              +--+
|  |  YES         |  | NO
+--+              +--+
```

If so, which type(s)? ______________________________

Why? ______________________________

What is your expected level of accuracy on this/these type(s) of projects at tender stage? ____________%

10. How important do you feel experience is in terms of forecasting design costs?

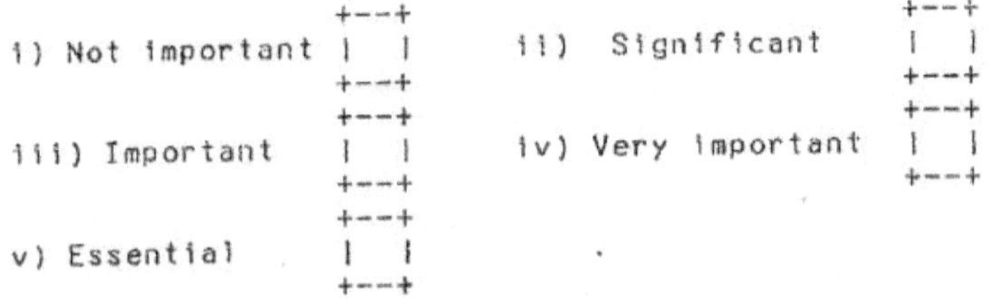

i) Not important [ ]   ii) Significant [ ]

iii) Important [ ]   iv) Very important [ ]

v) Essential [ ]

11. How do you rate your forecasting expertise?

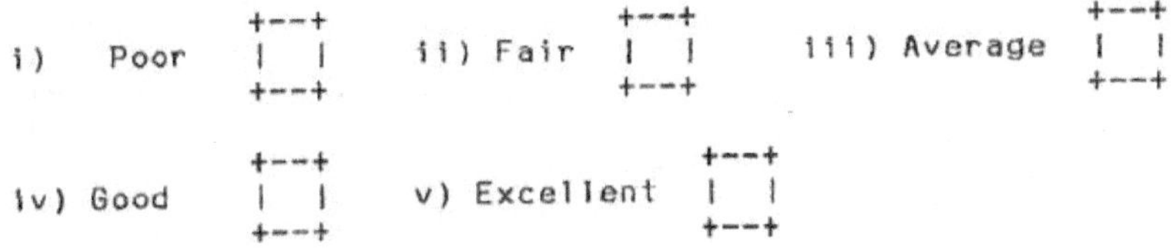

i) Poor [ ]   ii) Fair [ ]   iii) Average [ ]

iv) Good [ ]   v) Excellent [ ]

12. Please indicate in the table below how the following factors influence the accuracy of your forecasts?

(KEY : 1 = no influence on accuracy of forecasts

2 = little influence

3 = some influence

4 = significant influence

5 = essential to accuracy of forecasts)

FACTOR	1	2	3	4	5
Type of Project					
Design Information Available					
Historical Data Available					
Experience with Similar Projects					
Expertise in Forecasting					
Geographical Location					

Value/Size of Project					
Degree of Services					
Complexity of Project					
Present Market Conditions					
Expected Future Market Conditions					
Duration of Project					
Cost Limits					
Site Conditions					
The Design Team					

13. Please state any other factors that you feel affect the accuracy of your design forecasts. ____________________

____

____

____

14. Please indicate in the table below to what extent you possess the following characteristics or personality traits. The extent of your ability is rated on a scale of one to five for each characteristic.

(KEY : 1 = none
2 = little
3 = adequate
4 = significant
5 = exceptional

FACTOR	1	2	3	4	5
Ability to identify cost significant aspects					
Ability to cope with insufficient design details					
Ability to visualise the building					
Intuition					
Experience in forecasting					
Understanding and knowledge of implications of market condition					
Analytical ability					
Logical and systematic approach					
Memory of similar project details					
Personality factors					
Length of time spent in the profession					
Qualifications					

15. Please state any other characteristics or abilities that you feel you possess which affect the accuracy of design forecasts.

YOUR TIME AND EFFORT IN COMPLETING THIS QUESTIONNAIRE IS MUCH APPRECIATED

GEOGRAPHICAL LOCATION CODES FOR EMPIRICAL DATA

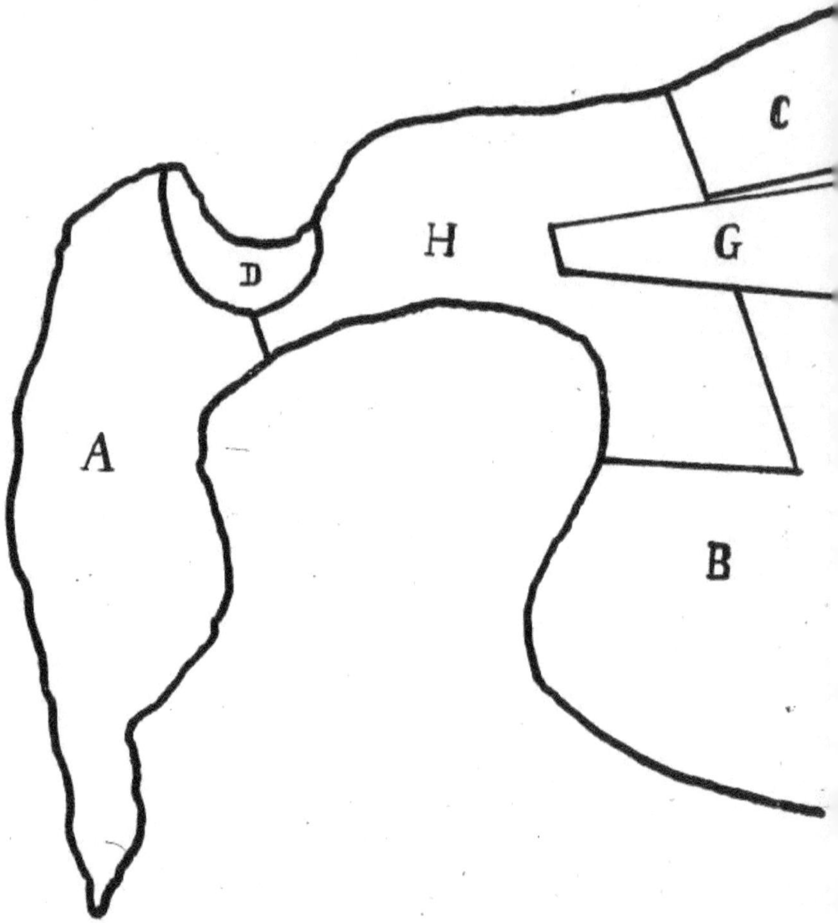

A = PENINSULA

B = BOLAND

C = ATLANTIS

D = METROPOLITAN

G = NORTHERN SUBURBS

H = CAPE FLATS

PERFORMANCE OF ESTIMATORS ON TENDERS
QUANTITIES [illegible] PWD IN 1990/91.

TENDERS ON ALL PROJECTS INCLUDING ALLOWANCES

DATE (YY/MM/DD)	AREA	[illegible]	[illegible]	SECTOR	CONTRACT	ITEMS	AMOUNT TENDERED	UPDATED VALUE	GROSS ESTIMATE	DEVIATION R's	[illegible] %	[illegible] %	[illegible]	NEXT TENDER	NEXT 3S ESTIMATE	NETT DEVIATION	[illegible] %	[illegible] %
[illegible]	Pretoria	[illegible]	[illegible]	PWD	PROJECT NO. 1	[illegible]	[illegible]	[illegible]	[illegible]	[illegible]	[illegible]	[illegible]	[illegible]	[illegible]	[illegible]	[illegible]	[illegible]	[illegible]
[illegible]	Port Elizabeth	[illegible]	[illegible]	PWD	PROJECT NO. 2	[illegible]	[illegible]	[illegible]	[illegible]	[illegible]	[illegible]	[illegible]	[illegible]	[illegible]	[illegible]	[illegible]	[illegible]	[illegible]
[illegible]	Pretoria	[illegible]	[illegible]	PWD	PROJECT NO. 3	[illegible]	[illegible]	[illegible]	[illegible]	[illegible]	[illegible]	[illegible]	[illegible]	[illegible]	[illegible]	[illegible]	[illegible]	[illegible]
[illegible]	Kempton Park	[illegible]	[illegible]	PWD	PROJECT NO. 4	[illegible]	[illegible]	[illegible]	[illegible]	[illegible]	[illegible]	[illegible]	[illegible]	[illegible]	[illegible]	[illegible]	[illegible]	[illegible]
[illegible]	Pretoria	[illegible]	[illegible]	PWD	PROJECT NO. 5	[illegible]	[illegible]	[illegible]	[illegible]	[illegible]	[illegible]	[illegible]	[illegible]	[illegible]	[illegible]	[illegible]	[illegible]	[illegible]
[illegible]	[illegible]	[illegible]	[illegible]	PWD	PROJECT NO. 6	[illegible]	[illegible]	[illegible]	[illegible]	[illegible]	[illegible]	[illegible]	[illegible]	[illegible]	[illegible]	[illegible]	[illegible]	[illegible]
[illegible]	Kimberley	[illegible]	[illegible]	PWD	PROJECT NO. 7	[illegible]	[illegible]	[illegible]	[illegible]	[illegible]	[illegible]	[illegible]	[illegible]	[illegible]	[illegible]	[illegible]	[illegible]	[illegible]
[illegible]	Pretoria	[illegible]	[illegible]	PWD	PROJECT NO. 8	[illegible]	[illegible]	[illegible]	[illegible]	[illegible]	[illegible]	[illegible]	[illegible]	[illegible]	[illegible]	[illegible]	[illegible]	[illegible]
[illegible]	Pretoria	[illegible]	[illegible]	PWD	PROJECT NO. 9	[illegible]	[illegible]	[illegible]	[illegible]	[illegible]	[illegible]	[illegible]	[illegible]	[illegible]	[illegible]	[illegible]	[illegible]	[illegible]
[illegible]	[illegible]	[illegible]	[illegible]	PWD	PROJECT NO. 10	[illegible]	[illegible]	[illegible]	[illegible]	[illegible]	[illegible]	[illegible]	[illegible]	[illegible]	[illegible]	[illegible]	[illegible]	[illegible]
[illegible]	Pretoria	[illegible]	[illegible]	PWD	PROJECT NO. 11	[illegible]	[illegible]	[illegible]	[illegible]	[illegible]	[illegible]	[illegible]	[illegible]	[illegible]	[illegible]	[illegible]	[illegible]	[illegible]
[illegible]	[illegible]	[illegible]	[illegible]	PWD	PROJECT NO. 12	[illegible]	[illegible]	[illegible]	[illegible]	[illegible]	[illegible]	[illegible]	[illegible]	[illegible]	[illegible]	[illegible]	[illegible]	[illegible]
[illegible]	Louis Trichardt	[illegible]	[illegible]	PWD	PROJECT NO. 13	[illegible]	[illegible]	[illegible]	[illegible]	[illegible]	[illegible]	[illegible]	[illegible]	[illegible]	[illegible]	[illegible]	[illegible]	[illegible]
[illegible]	Louis Trichardt	[illegible]	[illegible]	PWD	PROJECT NO. 14	[illegible]	[illegible]	[illegible]	[illegible]	[illegible]	[illegible]	[illegible]	[illegible]	[illegible]	[illegible]	[illegible]	[illegible]	[illegible]
[illegible]	Louis Trichardt	[illegible]	[illegible]	PWD	PROJECT NO. 15	[illegible]	[illegible]	[illegible]	[illegible]	[illegible]	[illegible]	[illegible]	[illegible]	[illegible]	[illegible]	[illegible]	[illegible]	[illegible]
[illegible]	[illegible]	[illegible]	[illegible]	PWD	PROJECT NO. 16	[illegible]	[illegible]	[illegible]	[illegible]	[illegible]	[illegible]	[illegible]	[illegible]	[illegible]	[illegible]	[illegible]	[illegible]	[illegible]
[illegible]	Cape Town	[illegible]	[illegible]	PWD	PROJECT NO. 17	[illegible]	[illegible]	[illegible]	[illegible]	[illegible]	[illegible]	[illegible]	[illegible]	[illegible]	[illegible]	[illegible]	[illegible]	[illegible]
[illegible]	Oudtshoorn	[illegible]	[illegible]	PWD	PROJECT NO. 18	[illegible]	[illegible]	[illegible]	[illegible]	[illegible]	[illegible]	[illegible]	[illegible]	[illegible]	[illegible]	[illegible]	[illegible]	[illegible]
[illegible]	[illegible]	[illegible]	[illegible]	PWD	PROJECT NO. 19	[illegible]	[illegible]	[illegible]	[illegible]	[illegible]	[illegible]	[illegible]	[illegible]	[illegible]	[illegible]	[illegible]	[illegible]	[illegible]
[illegible]	Cape Town	[illegible]	[illegible]	PWD	PROJECT NO. 20	[illegible]	[illegible]	[illegible]	[illegible]	[illegible]	[illegible]	[illegible]	[illegible]	[illegible]	[illegible]	[illegible]	[illegible]	[illegible]
[illegible]	Paarl	[illegible]	[illegible]	PWD	PROJECT NO. 21	[illegible]	[illegible]	[illegible]	[illegible]	[illegible]	[illegible]	[illegible]	[illegible]	[illegible]	[illegible]	[illegible]	[illegible]	[illegible]
[illegible]	Saldanha	[illegible]	[illegible]	PWD	PROJECT NO. 22	[illegible]	[illegible]	[illegible]	[illegible]	[illegible]	[illegible]	[illegible]	[illegible]	[illegible]	[illegible]	[illegible]	[illegible]	[illegible]
[illegible]	Ficksburg	[illegible]	[illegible]	PWD	PROJECT NO. 23	[illegible]	[illegible]	[illegible]	[illegible]	[illegible]	[illegible]	[illegible]	[illegible]	[illegible]	[illegible]	[illegible]	[illegible]	[illegible]
[illegible]	Rustenburg	[illegible]	[illegible]	PWD	PROJECT NO. 24	[illegible]	[illegible]	[illegible]	[illegible]	[illegible]	[illegible]	[illegible]	[illegible]	[illegible]	[illegible]	[illegible]	[illegible]	[illegible]
[illegible]	Pretoria	[illegible]	[illegible]	PWD	PROJECT NO. 25	[illegible]	[illegible]	[illegible]	[illegible]	[illegible]	[illegible]	[illegible]	[illegible]	[illegible]	[illegible]	[illegible]	[illegible]	[illegible]
[illegible]	Grahamstown	[illegible]	[illegible]	PWD	PROJECT NO. 26	[illegible]	[illegible]	[illegible]	[illegible]	[illegible]	[illegible]	[illegible]	[illegible]	[illegible]	[illegible]	[illegible]	[illegible]	[illegible]
[illegible]	Pietersburg	[illegible]	[illegible]	PWD	PROJECT NO. 27	[illegible]	[illegible]	[illegible]	[illegible]	[illegible]	[illegible]	[illegible]	[illegible]	[illegible]	[illegible]	[illegible]	[illegible]	[illegible]
[illegible]	Pretoria	[illegible]	[illegible]	PWD	PROJECT NO. 28	[illegible]	[illegible]	[illegible]	[illegible]	[illegible]	[illegible]	[illegible]	[illegible]	[illegible]	[illegible]	[illegible]	[illegible]	[illegible]
[illegible]	Bellville	[illegible]	[illegible]	PWD	PROJECT NO. 29	[illegible]	[illegible]	[illegible]	[illegible]	[illegible]	[illegible]	[illegible]	[illegible]	[illegible]	[illegible]	[illegible]	[illegible]	[illegible]
[illegible]	Cape Town	[illegible]	[illegible]	PWD	PROJECT NO. 30	[illegible]	[illegible]	[illegible]	[illegible]	[illegible]	[illegible]	[illegible]	[illegible]	[illegible]	[illegible]	[illegible]	[illegible]	[illegible]
[illegible]	Mossel Bay	[illegible]	[illegible]	PWD	PROJECT NO. 31	[illegible]	[illegible]	[illegible]	[illegible]	[illegible]	[illegible]	[illegible]	[illegible]	[illegible]	[illegible]	[illegible]	[illegible]	[illegible]
[illegible]	Parow	[illegible]	[illegible]	PWD	PROJECT NO. 32	[illegible]	[illegible]	[illegible]	[illegible]	[illegible]	[illegible]	[illegible]	[illegible]	[illegible]	[illegible]	[illegible]	[illegible]	[illegible]
[illegible]	[illegible]	[illegible]	[illegible]	PWD	PROJECT NO. 33	[illegible]	[illegible]	[illegible]	[illegible]	[illegible]	[illegible]	[illegible]	[illegible]	[illegible]	[illegible]	[illegible]	[illegible]	[illegible]
[illegible]	[illegible]	[illegible]	[illegible]	PWD	PROJECT NO. 34	[illegible]	[illegible]	[illegible]	[illegible]	[illegible]	[illegible]	[illegible]	[illegible]	[illegible]	[illegible]	[illegible]	[illegible]	[illegible]
[illegible]	Grahamstown	[illegible]	[illegible]	PWD	PROJECT NO. 35	[illegible]	[illegible]	[illegible]	[illegible]	[illegible]	[illegible]	[illegible]	[illegible]	[illegible]	[illegible]	[illegible]	[illegible]	[illegible]
[illegible]	Pretoria	[illegible]	[illegible]	PWD	PROJECT NO. 36	[illegible]	[illegible]	[illegible]	[illegible]	[illegible]	[illegible]	[illegible]	[illegible]	[illegible]	[illegible]	[illegible]	[illegible]	[illegible]
[illegible]	[illegible]	[illegible]	[illegible]	PWD	PROJECT NO. 37	[illegible]	[illegible]	[illegible]	[illegible]	[illegible]	[illegible]	[illegible]	[illegible]	[illegible]	[illegible]	[illegible]	[illegible]	[illegible]
[illegible]	Port Elizabeth	[illegible]	[illegible]	PWD	PROJECT NO. 38	[illegible]	[illegible]	[illegible]	[illegible]	[illegible]	[illegible]	[illegible]	[illegible]	[illegible]	[illegible]	[illegible]	[illegible]	[illegible]
[illegible]	Ellisras	[illegible]	[illegible]	PWD	PROJECT NO. 39	[illegible]	[illegible]	[illegible]	[illegible]	[illegible]	[illegible]	[illegible]	[illegible]	[illegible]	[illegible]	[illegible]	[illegible]	[illegible]
[illegible]	[illegible]	[illegible]	[illegible]	PWD	PROJECT NO. 40	[illegible]	[illegible]	[illegible]	[illegible]	[illegible]	[illegible]	[illegible]	[illegible]	[illegible]	[illegible]	[illegible]	[illegible]	[illegible]
[illegible]	[illegible]	[illegible]	[illegible]	PWD	PROJECT NO. 41	[illegible]	[illegible]	[illegible]	[illegible]	[illegible]	[illegible]	[illegible]	[illegible]	[illegible]	[illegible]	[illegible]	[illegible]	[illegible]
[illegible]	[illegible]	[illegible]	[illegible]	PWD	PROJECT NO. 42	[illegible]	[illegible]	[illegible]	[illegible]	[illegible]	[illegible]	[illegible]	[illegible]	[illegible]	[illegible]	[illegible]	[illegible]	[illegible]
[illegible]	Robben Island	[illegible]	[illegible]	PWD	PROJECT NO. 43	[illegible]	[illegible]	[illegible]	[illegible]	[illegible]	[illegible]	[illegible]	[illegible]	[illegible]	[illegible]	[illegible]	[illegible]	[illegible]
[illegible]	Louis Trichardt	[illegible]	[illegible]	PWD	PROJECT NO. 44	[illegible]	[illegible]	[illegible]	[illegible]	[illegible]	[illegible]	[illegible]	[illegible]	[illegible]	[illegible]	[illegible]	[illegible]	[illegible]
[illegible]	Louis Trichardt	[illegible]	[illegible]	PWD	PROJECT NO. 45	[illegible]	[illegible]	[illegible]	[illegible]	[illegible]	[illegible]	[illegible]	[illegible]	[illegible]	[illegible]	[illegible]	[illegible]	[illegible]
TOTAL					45 PROJECTS RECORDED	[illegible]	[illegible]	[illegible]	[illegible]	[illegible]			[illegible]	[illegible]	[illegible]	[illegible]		
MEAN						[illegible]	[illegible]	[illegible]	[illegible]	[illegible]			[illegible]	[illegible]	[illegible]	[illegible]		

MEASURES OF ACCURACY

RANGE (MAXIMUM)	[illegible]	[illegible]
RANGE (MINIMUM)	[illegible]	[illegible]
RANGE (MAXIMUM TO MINIMUM)	[illegible]	[illegible]
MEAN ERROR	[illegible]	[illegible]
MEAN DEVIATION	[illegible]	[illegible]
STANDARD DEVIATION	[illegible]	[illegible]
COEFFICIENT OF VARIATION	[illegible]	[illegible]

CONTRACT DETAILS							TENDER DETAILS						ESTIMATE DETAILS											
DATE	AREA	CA/ SUB	ALT/ NEW	SECTOR	INDEX BER	CONTRACT	BIDS	AMOUNT TENDERED	UPDATED VALUE	RANGE BID %	MEAN TENDER	CV	GROSS ESTIMATE	DEVIATION R's	GROSS % REL	GRABS % DEV	FIXED SUMS ETC	ALLOW %	TENDER NETT	ESTIMATE NETT	NETT DEVIATION	NTDEV % REL	NTABS % DEV	QS
87/01/30	5	0	N	PVT	338.80	ORPHANAGE	11	1363880	3099727	33.71	1523682	7.64	1510000	146120	10.71	10.71	221630	16.25	1142250	1288370	146120	12.79	12.79	21
87/02/03	1	3	N	PVT	345.30	SHOPS / RESTAURANT	12	510000	1137272	9.38	537431	2.77	525000	15000	2.94	2.94	56374	11.05	453626	468626	15000	3.31	3.31	9
87/02/17	5	8	N	PVT	345.30	RESIDENTIAL UNITS	20	303338	676427	64.70	376518	10.92	422000	118662	39.12	39.12	0	0.00	303338	422000	118662	39.12	39.12	21
87/02/17	4	7	N	PUB/1	345.30	UNIVERSITY SCIENCE BLDG	7	8964424	19990172	4.92	9146450	1.86	9100000	135576	1.51	1.51	2234894	24.93	6729530	6865106	135576	2.01	2.01	9
87/03/02	4	4	A	PVT	350.40	COLLEGE OF MEDICINE	9	777450	1700438	13.76	814793	4.64	835000	57550	7.40	7.40	187400	24.10	590050	647600	57550	9.75	9.75	9
87/03/04	3	4	A	PVT	357.40	LADIES HOME	6	896000	1968950	7.93	926761	2.32	866000	-30000	-3.35	3.35	196000	21.88	700000	670000	-30000	-4.29	4.29	34
87/03/12	8	0	N	PVT	350.40	OLD AGE HOME	10	315742	693839	50.58	366081	11.09	310000	-5742	-1.82	1.82	47000	14.89	268742	263000	-5742	-2.14	2.14	26
87/03/25	6	8	N	PUB/1	350.40	UNIVERSITY RESIDENCE	11	7755023	17093575	10.84	8018418	3.21	7800000	44977	0.58	0.58	1384779	17.86	6370244	6415221	44977	0.71	0.71	1
87/04/16	A	3	A	PVT	355.30	ALT. OFFICES	10	485584	1054737	19.83	538250	6.39	481600	-3984	-0.82	0.82	52000	10.71	433584	429600	-3984	-0.92	0.92	11
87/04/12	3	4	N	PVT	355.30	CRECHE	10	408223	894196	8.76	418848	2.36	424000	15777	3.86	3.86	42250	10.35	365973	381750	15777	4.31	4.31	9
87/04/22	N	3	N	PUB/1	355.50	COMMUNITY CENTRE	8	338147	752414	41.56	368589	11.42	345000	6853	2.03	2.03	27470	8.12	310677	317530	6853	2.21	2.21	9
87/04/29	4	7	N	PUB/1	355.50	SCHOOL	8	1636438	3594465	9.87	1724046	3.53	1850290	213852	13.07	13.07	65000	3.97	1571438	1785290	213852	13.61	13.61	29
87/05/27	6	7	N	PUB/1	360.60	TECHNIKON LIBRARY	11	3357000	7160363	8.73	3522740	3.00	3360000	3000	0.09	0.09	667725	19.89	2689275	2692275	3000	0.11	0.11	9
87/06/10	5	7	N	PUB/1	366.80	UNIVERSITY ADMIN BLOCK	13	1768668	3794156	14.24	1902436	3.93	1620000	-148668	-8.42	8.42	475405	26.94	1293263	1144595	-148668	-11.50	11.50	21
87/07/24	7	3	A	PVT	373.00	ALTS FOR PARKING	6	184481	380833	7.60	190331	2.32	190000	5519	2.99	2.99	41850	22.69	142631	148150	5519	3.87	3.87	9
87/09/02	A	8	N	PVT	384.20	HOUSE	6	783000	1568852	20.84	875515	5.37	800000	17000	2.17	2.17	384737	49.14	398263	415263	17000	4.27	4.27	9
87/09/25	A	7	N	PUB/2	384.30	LIBRARY	11	1244200	2492933	14.09	1317272	4.02	1306177	61977	4.98	4.98	317100	25.49	927100	989077	61977	6.69	6.69	8
87/10/20	A	7	N	PVT	387.30	UNIVERSITY TEACHING BLDG	7	2562631	5068691	4.79	2614930	1.37	2072000	309349	19.88	19.88	779467	30.42	1783084	2092533	309349	28.56	28.56	21
87/10/28	6	7	N	PUB/1	389.30	TECHNIKON ADMIN BLDG.	6	3976040	7864285	0.98	4216745	2.63	3990000	13960	0.35	0.35	707000	17.78	3269040	3283000	13960	0.43	0.43	9
87/11/16	N	6	N	PVT	394.40	CHURCH	6	642183	1253755	8.92	663668	7.98	655000	12817	2.00	2.00	91500	14.25	550683	563500	12817	2.33	2.33	34
87/11/18	N	8	N	PVT	394.40	2 BLOCKS FLATS	14	1365838	2666530	21.46	1453932	4.09	1500000	134162	9.82	9.82	241000	17.64	1124838	1259000	134162	11.93	11.93	6
87/11/30	N	4	N	PUB/1	394.40	TRAINING CENTRE	9	751000	1672059	11.26	804535	3.00	782000	31000	4.13	4.13	134300	17.88	616700	647700	31000	5.03	5.03	9
87/12/04	A	7	N	PVT	404.30	RESEARCH CENTRE	9	3560560	6777027	7.07	3710023	2.09	2850000	-710560	-19.96	19.96	640000	17.97	2920560	2210000	-710560	-24.33	24.33	17
87/12/30	3	7	N	PUB/1	404.50	TECHNIKON TEACHING BLDG.	6	12800102	24369805	4.27	13067483	1.58	20775219	7975117	62.28	62.28	4624269	36.12	8175833	16150950	7975117	97.50	97.50	21
87/12/12	6	3	N	PVT	404.50	OFFICES	6	2121212	4037967	8.24	2186368	2.74	2030000	-91212	-4.30	4.30	419000	19.75	1702212	1611000	-91212	-5.36	5.36	17
88/02/05	6	6	N	PVT	424.90	CHURCH RECREATION HALL	5	373373	676625	20.75	410287	11.98	387500	14127	3.78	3.78	64900	17.38	308473	322600	14127	4.58	4.58	3
88/02/10	N	2	N	PUB/1	424.90	TELEPHONE EXCHANGE	7	796600	1443591	29.40	901120	3.35	894000	97400	12.23	12.23	0	0.00	796600	894000	97400	12.23	12.23	1
88/02/16	4	2	A	PVT	424.90	SERVICE STATION	13	629750	1141227	26.43	730537	7.69	638000	8250	1.31	1.31	243200	38.62	386550	394800	8250	2.13	2.13	12
88/02/19	R	7	N	PVT	424.90	MUNICIPAL LIBRARY	3	2764764	5010280	7.78	2864324	8.51	1956000	-808764	-29.25	29.25	398400	14.41	2366364	1557600	-808764	-34.18	34.18	12
88/02/25	3	4	A	PVT	424.90	ALT/EXT HOSPITAL	5	3227227	5848332	11.16	3411421	6.49	3300000	72773	2.25	2.25	2000000	61.97	1227227	1300000	72773	5.93	5.93	7
88/03/01	A	8	A	PVT	429.80	HOUSE ALTERATIONS	11	878000	1572964	23.01	965160	13.36	948000	70000	7.97	7.97	419500	47.78	458500	528500	70000	15.27	15.27	11
88/03/10	A	2	A	PVT	429.80	FACTORY EXTENSIONS	8	5535535	9917082	12.80	6018035	6.73	5435000	-100535	-1.82	1.82	74312	1.34	5461223	5360688	-100535	-1.84	1.84	9
88/03/17	4	3	A	PVT	429.80	ALTS/ADDS OFFICES	6	836194	1498067	9.59	886089	4.20	790000	-46194	-5.52	5.52	203270	24.31	632924	586730	-46194	-7.30	7.30	24
88/03/30	4	2	N	PUB/1	429.80	TELEPHONE EXCHANGE	7	594495	1065056	10.81	633291	3.40	630000	35505	5.97	5.97	6000	1.01	588495	624000	35505	6.03	6.03	4
88/04/20	3	7	N	PVT	434.70	SCHOOL	6	789789	1398982	15.92	823596	7.49	600000	-189789	-24.03	24.03	0	0.00	789789	600000	-189789	-24.03	24.03	3
88/04/30	A	5	N	PVT	434.70	THEATRE	8	1485485	2631294	27.46	1629679	7.84	1250000	-235485	-15.85	15.85	0	0.00	1485485	1250000	-235485	-15.85	15.85	1
88/05/04	6	7	N	PUB/2	437.60	LIBRARY	10	6546933	11467538	18.99	7136790	3.90	6740000	193067	2.95	2.95	2766000	42.25	3780933	3974000	193067	5.11	5.11	17
88/06/03	3	2	A	PVT	446.40	ALT/ADDS FACTORY	5	1299980	2238099	14.49	1382045	2.49	1086598	-213382	-16.41	16.41	190000	14.62	1109980	896598	-213382	-19.22	19.22	14
88/06/23	6	2	N	PVT	446.40	FACTORY OFFICE WAREHOUSE	6	3768330	6500032	7.38	3874881	1.64	3930000	161670	4.29	4.29	1689500	44.83	2078830	2240500	161670	7.78	7.78	1
88/06/30	4	2	N	PVT	446.40	WAREHOUSE	5	2678000	4619310	4.67	2737138	6.06	2800000	122000	4.56	4.56	1650000	61.61	1028000	1150000	122000	11.87	11.87	6
88/07/14	A	0	N	PVT	453.20	HOTEL	5	7567774	12860706	6.59	7795384	2.33	7464449	-103325	-1.37	1.37	2850000	37.66	4717774	4614449	-103325	-2.19	2.19	8
88/07/16	5	7	N	PVT	453.10	LABORATORY	4	866866	1473156	5.62	896740	1.94	810000	-56866	-6.56	6.56	437000	50.41	429866	373000	-56866	-13.23	13.23	60
88/08/19	A	3	N	PVT	459.90	PARKING GARAGE	5	5454163	9138477	22.18	6102819	6.40	6634000	1179837	21.63	21.63	1323167	24.26	4130996	5310833	1179837	28.56	28.56	20
88/08/31	4	2	N	PVT	459.90	OFFICES	3	1846700	3091037	5.93	1914464	2.33	1846506	-194	-0.01	0.01	347100	18.80	1499600	1499406	-194	-0.01	0.01	12
88/09/06	3	2	A	PVT	470.20	ALT OFFICES WAREHOUSE	4	1188837	1946841	8.12	1257890	2.20	1300649	111812	9.41	9.41	120400	10.13	1068437	1180249	111812	10.46	10.46	8
88/09/16	4	2	N	PVT	470.20	TRAINING CENTRE	15	2434493	3986237	18.47	2596979	4.87	2456900	22407	0.92	0.92	522000	21.44	1912493	1934900	22407	1.17	1.17	13
88/09/30	4	7	N	PVT	470.20	RESEARCH CENTRE	9	964141	1578878	36.12	1155350	10.76	890000	-74141	-7.69	7.69	0	0.00	964141	890000	-74141	-7.69	7.69	17
88/10/14	4	5	N	PUB/1	480.60	RECREATION AREA	9	4549456	7288971	9.70	4844938	2.05	4600000	50544	1.11	1.11	2760567	60.68	1788889	1839433	50544	2.83	2.83	9
88/11/01	4	1	N	PVT	490.90	SERVICE STATION	7	746000	1170136	56.65	844685	26.05	752250	6250	0.84	0.84	75800	10.16	670200	676450	6250	0.93	0.93	13
88/11/15	4	7	N	PVT	490.90	SCHOOL	6	1291250	2025357	9.91	1329093	3.13	1250000	-41250	-3.19	3.19	226300	17.53	1064950	1023700	-41250	-3.87	3.87	54
88/11/22	4	7	N	PUB/1	490.90	SCHOOL	7	918918	1441367	19.92	1000739	3.99	896000	-22918	-2.49	2.49	68670	7.47	850248	827330	-22918	-2.70	2.70	9
88/11/23	4	1	N	PVT	490.90	SERVICE STATION	11	423750	664672	30.24	474858	9.18	401000	-22750	-5.37	5.37	141000	33.27	282750	260000	-22750	-8.05	8.05	13
88/11/23	N	0	N	PVT	490.90	HOSTEL	3	598000	937991	29.86	705844	7.48	665000	67000	11.20	11.20	32000	5.35	566000	633000	67000	11.84	11.84	19
88/11/25	4	0	N	PVT	490.90	FLATS	13	1813813	2845053	26.89	1946915	7.37	1800000	-13813	-0.76	0.76	405000	22.33	1408813	1395000	-13813	-0.98	0.98	6
88/11/30	N	0	N	PVT	490.90	SENIORS CENTRE	4	616597	967162	3.61	630890	1.37	615000	-1597	-0.26	0.26	117000	18.98	499597	498000	-1597	-0.32	0.32	19
88/12/06	4	4	A	PUB/1	501.80	EXT HOSPITAL	5	3040000	4674121	4.37	3094683	1.46	2900000	-140000	-4.61	4.61	981350	32.28	2058650	1918650	-140000	-6.80	6.80	11
88/12/12	A	0	N	PVT	501.80	OLD AGE HOME	9	478660	735959	19.27	523422	5.03	485000	6340	1.32	1.32	69795	14.58	408865	415205	6340	1.55	1.55	9
88/12/14	4	3	N	PUB/2	501.80	TOWN HALL	10	3434340	5280415	21.07	3747645	5.63	3390000	-44340	-1.29	1.29	807000	23.50	2627340	2583000	-44340	-1.69	1.69	17

TENDERS ON ALL PROJECTS WITH QS ESTIMATES PROVIDED.

CONTRACT DETAILS							TENDER DETAILS						ESTIMATE DETAILS											
DATE	AREA	CY/SIR	ALT/NEW	SECTOR	INDEX BER	CONTRACT	BIDS	AMOUNT TENDERED	UPDATED VALUE	RANGE BID %	MEAN TENDER	CV	GROSS ESTIMATE	DEVIATION R's	GRDEV % REL	GRABS % DEV	FIXED SUMS ETC	ALLOW %	TENDER NETT	ESTIMATE NETT	NETT DEVIATION	NTDEV % REL	NABS % DEV	QS
89/01/19	W	3	4	PVT	310.60	FACTORY	6	696184	1049866	16.31	758538	5.78	680000	-16184	-2.32	2.32	95500	13.72	600624	584500	-16084	-2.69	2.69	1
89/02/17	A	4	4	PVT	320.20	SCHOOL	10	3069017	4540143	9.33	3192761	3.30	3120000	50981	1.66	1.66	860636	28.01	2208383	2259364	50981	2.31	2.31	20
89/03/08	A	5	4	PVT	329.00	SPORTS PAVILION	8	629060	914175	11.03	666772	3.89	621000	-9500	-1.43	1.43	121390	19.30	397599	478100	-9000	-1.77	1.77	9
89/03/20	D	2	A	PVT	329.80	REFURBISHM'T OFFICES	6	1689525	2455520	20.44	1824580	7.01	1700000	10475	0.62	0.62	802000	47.35	889575	900000	10475	1.18	1.18	7
89/04/21	W	4	4	PVT	339.20	ALT TO ORPHANAGE	8	697185	993669	17.33	757321	5.63	725000	31815	4.36	4.36	122000	17.50	575185	607000	31815	5.53	5.53	9
89/05/16	A	3	4	PVT	348.30	ALT TO GOLF CLUB	10	502215	719062	14.53	553250	4.34	500000	-4215	-0.83	0.83	161200	31.47	351015	346800	-4215	-1.20	1.20	3
89/05/26	W	7	4	PVT	348.30	PREPARATORY SCHOOL	7	729886	1024623	20.30	816409	6.22	851400	121514	16.65	16.65	92650	12.69	637236	758750	121514	19.07	19.07	9
89/06/07	W	1	4	PVT	348.90	SERVICE STATION	5	412986	574531	12.94	430990	4.92	410000	5029	1.22	1.22	120250	29.12	292730	297750	5020	1.71	1.71	12
89/06/07	W	1	4	PVT	348.90	SERVICE STATION	5	452364	578467	15.23	441959	5.78	423000	14636	3.55	3.55	115500	27.89	297364	312000	14636	4.92	4.92	12
89/06/08	A	3	4	PVT	348.90	OFFICES	14	945706	1326633	20.58	1056095	7.93	994300	48600	5.14	5.14	350000	37.03	594820	643420	48600	8.17	8.17	20
89/06/20	W	7	4	PVT	348.90	TECHNIKON BLDGS	7	5232536	7312175	11.79	5831128	3.66	4920000	-292536	-5.61	5.61	880000	15.08	4332536	4040000	-292536	-6.75	6.75	6
89/06/21	G	7	4	PUB/1	348.90	PRIMARY SCHOOL	6	2292422	3266524	6.68	2383942	3.40	2603000	192078	8.38	8.38	801941	35.28	1484981	1677059	192078	12.93	12.93	21
89/06/27	W	3	4	PVT	348.90	TURF CLUB PAVILION	4	9825000	13796583	2.96	10004491	1.09	9819001	-15199	-0.15	0.15	5202839	52.90	4632111	4616972	-15199	-0.33	0.33	8
89/07/21	A	2	4	PVT	349.30	EXT TO FACTORY	6	1055904	1476079	6.69	1391948	2.49	1150000	97000	9.21	9.21	230000	21.84	823000	920000	97000	11.79	11.79	6
89/07/25	A	3	4	PVT	369.30	FACTORY	7	2829940	3966965	11.39	2973697	3.75	2780000	-49940	-1.76	1.76	1048230	37.04	1701310	1731370	-49940	-2.00	2.00	22
89/07/28	W	8	4	PVT	399.30	SCHOOL FOR THE DEAF	4	1021904	1431222	10.68	1094421	3.92	980000	-41000	-4.02	4.02	36000	4.93	971000	930000	-41000	-4.22	4.22	17
89/08/01	A	7	4	PUB/1	349.80	TRAINING CENTRE	7	3529886	4943638	9.78	3707534	2.89	3400000	-129886	-3.68	3.68	589500	16.69	2940886	2811000	-129886	-4.42	4.42	52
89/08/22	W	7	4	PUB/1	349.80	UNIVERSITY LABORATORIES	6	7689967	10781074	3.88	7845298	1.48	8442567	745000	9.68	9.68	5107600	66.38	2584367	3329367	745000	28.77	28.72	30
89/08/25	A	3	4	PVT	349.80	RESTAURANT	6	1785433	2500525	14.75	1934663	5.37	1878527	93094	5.21	5.21	766000	42.90	1019433	1112527	93094	9.13	9.13	63
89/09/19	W	2	4	PVT	359.10	FACTORY / WAREHOUSE	11	3296309	4470860	30.12	3896325	6.76	3700000	453691	13.98	13.98	785000	24.18	2461309	2915000	453691	18.43	18.43	61
89/10/06	W	3	4	PVT	368.40	SHOPPING CENTRE	6	863806	1170287	14.13	939407	4.32	700000	-163806	-18.97	18.97	264000	30.56	599806	436000	-163806	-27.32	27.32	6
89/10/20	A	7	4	PVT	368.40	PRIMARY SCHOOL	7	610028	826380	5.85	632557	1.84	560140	-49920	-8.16	8.16	30450	4.99	579570	529650	-49920	-8.61	8.61	20
89/10/25	A	4	4	PUB/1	368.40	RENOVATIONS TO OFFICES	7	2896177	3923592	23.64	3379482	7.81	2822000	-74177	-2.56	2.56	593000	20.61	2299177	2225000	-74177	-3.23	3.23	5
89/10/20	A	7	4	PUB/1	368.40	EXT TO SCHOOL	3	733990	994331	19.73	801146	5.82	773900	38010	5.18	5.18	36030	4.91	697960	735970	38010	5.45	5.45	20
89/11/01	G	7	4	PUB/2	377.70	WORKSHOPS AT TECH COLLEGE	9	490000	663770	20.72	571949	8.19	500000	2000	0.40	0.40	28000	5.67	470000	472000	2000	0.43	0.43	19
89/11/03	W	5	4	PVT	377.70	OFFICE BLOCK	8	1669272	2274925	17.38	1769254	4.88	1701000	31728	1.90	1.90	461500	27.65	1207772	1239500	31728	2.63	2.63	35
89/11/22	C	5	4	PVT	377.70	RESORT DEVELOPMENT	7	3803243	5069255	20.52	4215752	5.67	3800000	-3243	-0.09	0.09	50000	1.31	3753243	3750000	-3243	-0.09	0.09	7
89/12/02	G	4	4	PVT	385.60	MEDICAL RESEARCH CENTRE	14	1824600	2401630	7.91	1899283	2.60	1564600	-260660	-14.28	14.28	641500	35.16	1183100	922500	-260600	-22.03	22.03	52
89/12/12	A	4	4	PVT	385.60	DAY CARE CENTRE	7	948400	1248321	40.06	1063001	9.38	950000	1600	0.17	0.17	100500	10.60	847900	849500	1600	0.19	0.19	9
90/01/26	A	8	4	PVT	392.40	TOWNHOUSE SCHEME	5	1569200	2029642	23.42	1798544	7.00	1793000	225800	14.39	14.39	356695	22.73	1212515	1438315	225800	18.62	18.62	14
90/02/07	A	8	4	PVT	399.70	HOUSE	4	672676	889379	5.17	711963	1.82	627000	-70676	-10.28	10.28	193000	27.59	501676	431000	-70676	-14.09	14.09	12
90/02/09	W	5	K	PUB/1	399.70	COMMUNITY CENTRE	8	1250625	1734167	10.32	1425690	3.21	1360000	9377	0.69	0.69	250000	18.38	1100623	1110000	9377	0.85	0.85	17
90/02/14	G	6	K	PVT	399.70	HOUSING	8	5319420	6830004	12.61	5708213	4.21	5373300	53880	1.01	1.01	819000	15.49	4500420	4554300	53880	1.20	1.20	52
90/02/19	W	7	K	PVT	399.70	OFFICES	9	2094934	2609844	17.02	2299177	5.10	2037473	-57439	-2.74	2.74	527500	25.18	1567434	1509973	-57459	-3.67	3.67	11
90/03/20	A	2	K	PVT	399.70	BREWERY DEPOT	5	5128018	6581657	14.29	5509004	5.29	5003800	-126000	-2.46	2.46	1311620	25.59	3814200	3688200	-126000	-3.30	3.30	502
90/03/22	W	3	K	PVT	407.80	SHOPPING CENTRE	3	30395134	38162476	0.74	30534971	0.37	31070000	11484866	28.44	28.44	13880058	34.30	26499076	37983942	11484866	43.34	43.34	20
90/04/09	W	6	K	PVT	415.80	OLD AGE HOME	3	4287000	5360490	36.35	5093896	7.23	4022844	-264156	-6.16	6.16	0	0.00	4287000	4022844	-264156	-6.16	6.16	10
90/04/09	D	1	A	PVT	415.90	ALT TO OFFICES	7	2209000	2762147	25.60	2522432	6.10	2250000	42600	1.86	1.86	1027000	46.78	1180000	1221900	41000	3.47	3.47	32
90/04/20	W	2	K	PVT	415.80	FACTORY	9	4549301	5688433	7.22	4727952	2.37	4500000	-49301	-1.08	1.08	847000	18.69	3701301	3652200	-49301	-1.33	1.33	502
90/04/27	A	3	K	PVT	425.00	BANK	12	2298767	2874539	15.58	2384940	3.65	2392000	93283	4.06	4.06	821000	35.72	1477717	1571000	93283	6.31	6.31	77
90/05/11	W	4	K	PVT	425.93	NURSERY AND GATEHOUSE	6	793295	991536	10.93	848125	3.70	797657	2362	0.30	0.30	0	0.00	795295	797657	2362	0.30	0.30	33
90/05/21	A	8	K	PVT	423.90	HOUSE	6	1670333	2071532	15.33	1809049	6.36	1650000	-28133	-1.69	1.69	485000	28.93	1193333	1165000	-28133	-2.37	2.37	13
90/05/23	W	4	K	PVT	423.90	CRECHE	8	447447	552237	33.69	517880	10.10	430110	-17337	-3.87	3.87	4000	0.89	443447	426110	-17337	-3.91	3.91	14
90/05/30	A	8	K	PVT	423.90	COTTAGES FOR THE AGED	7	2282590	2817109	14.21	2449866	3.16	2400000	117410	5.14	5.14	122000	5.34	2160590	2278000	117410	5.43	5.43	19
90/06/06	A	1	K	PVT	630.60	SERVICE STATION	8	720000	879163	25.85	790253	6.20	760000	40000	5.56	5.56	57500	7.99	662500	702500	40000	6.04	6.04	12
90/06/06	G	7	K	PVT	630.60	UNIVERSITY TEACHING BLDG.	6	3032389	3702624	13.88	3307857	4.40	3032000	-385	-0.01	0.01	335000	11.05	2697389	2697000	-385	-0.01	0.01	6
90/06/15	D	3	A	PVT	630.60	ALT. TO OFFICES	8	1932000	2359087	16.66	2054251	3.44	2209490	277490	14.36	14.36	1383700	71.57	549300	826790	277490	50.52	50.52	33
90/06/20	A	4	K	PUB/1	630.60	SECURITY OFFICE AND WARD	12	2620000	3199125	20.81	2989509	4.42	3000000	380000	14.50	14.50	160700	6.13	2459300	2839300	380000	15.45	15.45	52
90/06/23	G	5	K	PVT	630.60	SPORTS CLUB BUILDING	5	1210600	1478214	6.33	1246320	2.04	1150000	-60600	-5.01	5.01	356000	29.41	854600	794000	-60600	-7.09	7.09	17
90/07/12	M	4	K	PVT	637.20	TRAINING CENTRE	5	486000	587288	16.32	532037	6.32	490000	4000	0.82	0.82	25200	5.19	460800	464800	4000	0.87	0.87	9
90/07/18	B	7	K	PUB/1	637.30	PRIMARY SCHOOL	7	786009	999833	32.03	876861	14.10	729635	-56374	-7.17	7.17	0	0.00	786009	729635	-56374	-7.17	7.17	33
90/07/27	A	3	K	PVT	637.20	HEALTH CARE CENTRE	9	501581	702707	16.14	620630	4.15	600000	18417	3.11	3.17	79000	13.58	502381	521000	18417	3.66	3.66	12
90/07/23	A	3	K	PVT	637.20	OFFICE BLOCK	3	19484967	23545864	3.12	25609222	1.15	19532602	-152765	-0.78	0.78	10446972	53.77	9018045	8865680	-152765	-1.69	1.69	8
90/07/27	E	3	K	PVT	637.20	OFFICE BLOCK	4	1877957	2269345	4.82	2174352	1.15	1278143	-99614	-5.30	5.30	1008800	53.73	869157	769543	-99614	-11.46	11.46	8
90/08/02	D	3	K	PVT	645.70	HOUSE	3	395313	715339	12.74	455015	4.75	382600	-14035	-2.43	2.43	86100	14.45	310415	493700	-14315	-2.84	2.84	81
90/08/08	A	4	4	PVT	643.90	OLD AGE HOME	3	433990	521260	17.35	480366	3.76	461700	25089	3.94	3.94	87300	20.07	346690	374290	25088	7.43	7.43	14
90/08/09	D	3	A	PVT	643.70	OFFICES	4	467202	598794	21.67	540263	6.11	441670	-25604	-5.48	5.48	0	0.00	467202	441678	-25604	-5.40	5.48	1

TENDERS ON ALL PROJECTS WITH QS ESTIMATES PROVIDED.

CONTRACT DETAILS							TENDER DETAILS						ESTIMATE DETAILS											
DATE	AREA	CS/SIZE	ALT/NEW	SECTOR	INDEX NUMBER	CONTRACT	BIDS	AMOUNT TENDERED	UPDATED VALUE	RANGE BID %	MEAN TENDER	CV	GROSS ESTIMATE	DEVIATION R's	GRDEV % REL	GRABS % DEV	FIXED SUMS ETC	ALLOW %	TENDER NETT	ESTIMATE NETT	NETT DEVIATION	NTDEV % REL	NTABS % DEV	OD
90/08/16	A	8	A	PVT	645.90	ALT TO HOUSE	5	3100000	3707697	9.38	3248891	3.31	2860000	-240000	-7.74	7.74	1279000	41.26	1821000	1581000	-240000	-13.18	13.18	24
90/08/17	A	8	A	PVT	641.90	ALT TO HOUSE	4	609706	729349	18.77	676000	6.18	605028	-4070	-0.67	0.67	124000	22.30	433906	449870	-4078	-0.90	0.90	14
90/08/22	G	2	A	PVT	645.90	ALT TO FACTORY	4	616793	737184	15.66	668193	5.15	650000	33207	5.38	5.38	75000	12.16	541793	575050	33263	6.13	6.13	7
90/08/31	D	3	A	PVT	641.90	OFFICES - ADDITIONAL FLOOR	5	339300	405748	8.39	353809	2.69	415000	75700	22.31	22.31	0	0.00	339300	415000	75700	22.31	22.31	3
90/09/04	G	2	N	PUB/2	647.50	BANQUETING HALL	8	1139200	1350843	10.77	1208549	3.65	980000	-159500	-13.98	13.98	388000	34.06	751300	592000	-159300	-21.20	21.20	17
90/09/24	A	8	N	PVT	647.30	HOUSING FOR AGED	7	3419783	4066769	42.72	4100398	13.27	3200000	-219783	-6.43	6.43	0	0.00	3419783	3200000	-219783	-6.43	6.43	19
90/09/26	A	8	A	PVT	647.50	EXT TO APARTMENT BLOCK	5	7765580	9234744	3.68	7920087	1.22	7325000	-440580	-5.67	5.67	0	0.00	7765580	7325000	-440580	-5.67	5.67	27
90/10/15	G	3	N	PVT	651.00	OFFICE BLOCK	16	5995000	7098860	6.95	6226136	1.99	6599750	604750	10.09	10.09	2759736	46.03	3235264	3840014	604750	18.69	18.69	3
90/10/15	D	3	N	PVT	651.00	SHOPS, OFFICES & CLINIC	9	5523000	6532381	14.03	5961627	3.88	5800000	277000	5.02	5.02	3209000	58.10	2314000	2591000	277000	11.97	11.97	13
90/10/18	H	3	N	PVT	651.00	HALL AND COTTAGE	10	2736935	3237225	32.47	3067403	12.90	2744084	7149	0.26	0.26	295700	10.80	2441235	2448384	7149	0.29	0.29	11
90/10/11	G	5	A	PUB/2	654.00	EXT TO MUNICIPAL OFFICES	11	3185000	3767284	19.31	3386476	5.22	3200000	15000	0.47	0.47	1111847	34.91	2073153	2088153	15000	0.72	0.72	52
90/10/23	S	6	N	PVT	651.00	CHURCH	9	1152075	1362663	22.67	1272251	7.75	1735000	582920	50.60	50.60	335543	27.39	816525	1419415	582920	69.68	69.68	21
90/11/02	H	2	N	PVT	654.60	OFFICES AND WAREHOUSE	5	675000	793986	9.93	714644	2.65	675000	0	0.00	0.00	158100	23.42	516900	516900	0	0.00	0.00	22
90/11/07	E	3	N	PUB/1	654.60	POLICE STATION	4	795000	935151	26.69	924546	8.05	863000	70000	8.81	8.81	4575	0.58	790425	860425	70000	8.85	8.85	12
90/11/23	G	3	A	PVT	654.60	SECOND FLOOR TO BUILDING	11	993999	1169232	15.31	1061715	4.13	907291	-86700	-8.72	8.72	311000	31.29	682999	596291	-86708	-12.70	12.70	33
90/11/29	A	1	N	PUB/2	654.60	MUNICIPAL STORAGE DEPOT	8	1543210	1815264	74.27	1662669	6.28	1579000	35790	2.32	2.32	0	0.00	1543210	1579000	35790	2.32	2.32	1
90/12/13	D	1	A	PVT	664.40	REDEVELOPMENT OF GARAGE	9	546800	633709	16.87	595561	5.67	577000	30200	5.52	5.52	87000	15.91	459800	490000	30200	6.57	6.57	12
91/01/25	A	3	A	PVT	674.10	ALT TO CANTEEN	5	1111111	1269182	10.34	1160576	3.56	1050000	-61111	-5.50	5.50	250000	20.70	861111	800000	-61111	-6.94	6.94	17
91/01/25	A	8	N	PVT	674.10	FLATS AND GARAGES	5	2739000	3128660	11.79	2956624	3.94	2925000	186000	6.79	6.79	811000	29.61	1928000	2114000	186000	9.65	9.65	6
91/01/25	F	2	A	PVT	674.10	ADDITIONS TO WAREHOUSE	6	1126000	1286189	16.51	1220812	4.09	1170000	44000	3.91	3.91	48000	4.26	1078000	1122000	44000	4.08	4.08	1
91/01/28	D	3	A	PVT	674.10	ALT TO MOTOR SHOWROOM	9	1447500	1653427	17.23	1597186	4.53	1600000	152500	10.54	10.54	330645	22.84	1116855	1269355	152500	13.65	13.65	53
91/01/30	A	3	A	PVT	674.10	ALT TO OLD AGE HOME	10	799300	913011	67.65	1023782	18.66	810000	10700	1.34	1.34	0	0.00	799300	810000	10700	1.34	1.34	24
91/02/08	G	2	N	PVT	685.90	FACTORY / WAREHOUSE	19	2295000	2583938	29.70	2620979	7.51	2850000	555000	24.18	24.18	0	0.00	2295000	2850000	555000	24.18	24.18	17
91/02/13	H	2	N	PVT	685.90	ATAI FACTORIES	8	3839393	4433345	10.11	4153787	2.57	4180000	340607	8.87	8.87	1857000	41.14	2082393	2323000	240607	11.55	11.55	30
91/02/20	D	8	N	PVT	685.90	NEW APARTMENT BUILDING	5	2587852	2913451	6.23	2676345	2.05	2549236	-38616	-1.49	1.49	901610	34.84	1686242	1647626	-38616	-2.29	2.29	22
91/02/22	D	3	N	PVT	685.90	OFFICES	6	1933331	2176304	8.96	2012227	3.14	2099180	165849	8.58	8.58	207500	10.73	1725831	1891500	165669	9.61	9.61	19
91/03/05	H	2	N	PVT	694.60	WORKSHOP	5	818000	908795	31.42	961354	10.61	950000	132000	16.14	16.14	35000	4.28	783000	915000	132000	16.86	16.86	19
91/03/15	H	4	N	PUB/1	694.60	HOSPITAL	14	1741296	1930534	21.46	1933603	5.15	2253000	511704	29.39	29.39	82000	4.71	1659296	2171000	511704	30.84	30.84	1
91/03/19	A	8	N	PVT	694.60	11 FLATS	6	6741147	7472900	10.65	7100113	3.40	7070380	329233	4.88	4.88	1591500	23.59	5110647	5479880	329233	6.39	6.39	14
91/04/09	D	3	A	PVT	705.40	REFURB OF OFFICES	9	549000	599277	22.50	612458	6.58	575000	26000	4.74	4.74	278250	50.68	270750	296750	26000	9.60	9.60	8
91/04/19	A	4	N	PVT	705.40	HOSPITAL	9	14362450	15677761	9.63	15157158	2.64	14962450	600000	4.18	4.18	0	0.00	14362450	14962450	600000	4.18	4.18	3
91/04/24	A	3	A	PVT	705.40	REDEVELOPMENT FOR BANK	9	7955297	8683037	6.20	8232565	1.63	7995000	39703	0.45	0.45	4817000	60.55	3138297	3178000	39703	1.14	1.14	3
91/05/03	A	7	N	PUB/2	716.10	MUNICIPAL LIBRARY	12	1078956	1160168	24.00	1194378	5.72	1052000	-26956	-2.50	2.50	28300	2.62	1050656	1023700	-26956	-2.57	2.57	1
91/05/08	H	7	A	PUB/1	716.10	ALT TO SCHOOL	4	591330	655039	14.28	613068	3.94	610440	19110	3.23	3.23	4000	0.68	587330	606440	19110	3.25	3.25	61
91/05/16	D	3	A	PVT	716.10	ALT TO OFFICES	7	2861339	3076709	0.27	2993276	2.05	2717000	-144339	-5.04	5.04	2020000	70.60	841339	697000	-144339	-17.16	17.16	1
91/05/24	A	6	A	PVT	716.10	ALT TO CREMATORIUM	7	757845	814007	9.21	803054	2.84	784724	26879	3.55	3.55	188000	24.81	569845	596724	26879	4.72	4.72	2
91/05/24	A	3	N	PVT	716.10	SHOPS AND OFFICES	6	1327138	1427030	17.71	1439769	5.51	1400000	72862	5.49	5.49	119000	8.97	1208138	1281000	72862	6.03	6.03	27
91/06/03	D	3	A	PVT	718.96	REFURBISHMENT OF ARCADE	5	6420268	6876626	12.22	6609620	4.21	8061000	1640732	25.56	25.56	5303129	82.60	1117139	2757871	1640732	146.87	146.87	23
91/06/04	A	3	A	PVT	718.96	ALT TO HOTEL	7	649000	695631	14.79	683821	5.26	665031	16031	2.47	2.47	388400	59.85	260600	276631	16031	6.15	6.15	33
91/06/12	D	7	A	PUB/1	718.96	RENOVATIONS TO ROOF	3	1746000	1870107	129.02	2500904	41.99	1061720	-684280	-39.22	39.22	0	0.00	1746000	1061720	-684280	-39.22	39.22	14
91/06/12	G	3	A	PVT	718.96	RENEW TO OFFICES	9	3108886	3329860	8.95	3241511	2.64	2658886	-450000	-14.47	14.47	1541578	49.59	1567308	1117308	-450000	-28.71	28.71	52
91/06/14	H	3	A	PVT	718.96	ALT TO FACTORY / WAREHOUSE	8	1593000	1706253	21.50	1675718	6.72	1300000	-293000	-18.39	18.39	929000	58.32	664000	371000	-293000	-44.13	44.13	302
91/06/19	A	3	A	PVT	718.96	ALT TO OFFICES	7	826113	884834	24.08	900063	6.43	854853	28740	3.48	3.48	270700	32.77	555413	584153	28740	5.17	5.17	33
91/06/21	A	3	N	PVT	718.96	SHOPPING CENTRE	12	5086000	5661253	9.19	5520119	2.64	4490000	-596000	-11.20	11.20	991975	18.71	4094025	3498025	-596000	-17.88	17.88	52
91/06/28	E	8	A	PVT	718.96	REFURBISHMENT OF HOTEL	4	336633	360561	24.03	382311	8.41	326894	-9739	-2.89	2.89	0	0.00	336633	326894	-9739	-2.89	2.89	3
91/07/10	H	7	N	PUB/2	721.80	PRIMARY SCHOOL	12	3673000	3918274	21.57	3981343	6.34	4100000	427000	11.63	11.63	6000	0.16	3667000	4094000	427000	11.64	11.64	19
91/07/19	A	8	N	PVT	721.80	HOUSE	4	1545201	1648884	16.91	1673312	5.24	1950000	404799	26.20	26.20	481500	31.16	1063701	1468500	404799	38.06	38.06	12
91/07/22	A	7	N	PVT	721.80	FACTORY	13	2133133	2275578	21.80	2314664	5.38	2380000	246867	11.57	11.57	350000	16.41	1783133	2030000	246867	13.84	13.84	52
91/07/29	E	5	A	PVT	721.80	REFURB TO RESTAURANT	3	598598	638571	15.47	645531	5.86	780000	181402	30.30	30.30	226850	37.90	371748	553150	181402	48.80	48.80	24
91/08/02	G	3	A	PVT	724.60	ADD. TO DISTRIBUTION CENTRE	6	3192000	3391996	6.56	3310436	1.99	3354776	162776	5.10	5.10	985805	30.88	2206195	2368971	162776	7.38	7.38	20
91/08/07	H	7	N	PUB/1	724.60	UNIVERSITY TEACHING BLDG.	10	1574711	1673375	20.59	1750053	5.26	1874000	299289	19.01	19.01	99045	6.29	1475666	1774955	299289	20.28	20.28	21
91/08/12	A	4	A	PVT	724.60	UNIVERSITY MEDICAL SCHOOL	3	390903	415393	24.19	443996	9.00	430000	39097	10.00	10.00	72750	18.61	318153	357250	39097	12.29	12.29	9
91/08/14	A	8	N	PUB/1	724.60	PRISON HOUSING	5	17227000	18306362	115.58	19935538	45.18	21000000	3773000	21.90	21.90	829000	4.81	16398000	20171000	3773000	23.01	23.01	9
91/08/16	G	3	N	PVT	724.60	OFFICE DEVELOPMENT	10	6042677	6421495	8.64	6292687	2.95	5950000	-92677	-1.54	1.54	2060000	34.09	3982677	3890000	-92677	-2.33	2.33	17
91/08/19	H	3	A	PVT	724.60	ALTS/EXT. TO COMPUTER BLDG	4	1507877	1602353	4.72	1534736	1.78	1468000	-39877	-2.64	2.64	867500	57.53	640377	600500	-39877	-6.23	6.23	25
91/08/22	G	5	N	PVT	724.60	OFFICES	9	11795000	12534017	11.09	12516944	3.71	12224000	429000	3.72	3.72	2851000	24.17	8944000	9373000	429000	4.71	4.71	1
91/08/27	A	3	A	PVT	724.60	ALTS TO OFFICE COMPLEX	7	9991400	10616989	4.90	10412718	5.33	10714536	723136	7.24	7.24	3727840	37.31	6263560	6986796	723236	11.55	11.55	33

TENDERS ON ALL PROJECTS WITH QS ESTIMATES PROVIDED.

DATE	AREA	CI/SIR	ALT/NEW	SECTOR	INDEX BER	CONTRACT	BIDS	AMOUNT TENDERED	UPDATED VALUE	RANGE BID %	MEAN TENDER	CV
CONTRACT DETAILS							TENDER DETAILS					
91/08/30	D	2	N	PUB/1	724.60	COLD STORAGE FACILITY	6	6928500	7367076	21.36	7415757	8.07
91/09/20	D	1	N	PVT	725.00	OFFICES	10	3809867	4044342	7.62	3961793	2.25
91/09/10	N	2	A	PVT	725.00	ALTS TO FACTORY	8	1391500	1477869	15.63	1494125	5.29
91/09/27	N	1	N	PVT	725.00	MOTOR SHOWROOM	9	1035000	1099241	39.84	1215974	10.90
91/09/30	N	6	A	PVT	725.00	DOCTORS SUITE & PHARMACY	5	718000	762566	24.40	765591	8.49
91/10/01	A	8	N	PVT	745.00	SECTIONAL TITLE HOUSING	6	1074000	1118040	23.56	1101600	7.71
91/10/08	A	3	N	PVT	745.00	HQ FOR WELFARE BODY	6	9346200	9639831	7.87	9620768	2.37
91/10/08	N	5	N	PVT	745.00	SHOPPING CENTRE	8	4705909	4863025	10.18	4979659	3.22
91/10/09	A	4	N	PUB/1	745.00	CARE & REHAB. CENTRE	4	7521460	7763522	17.44	7827969	5.80
91/10/25	A	7	A	PVT	745.00	ALTS/ADDS TO PRIMARY SCHOOL	10	998471	1031977	128.36	1719607	22.20
91/10/31	A	6	N	PVT	745.00	RELIGIOUS MEETING HOUSE	9	736868	763595	55.23	820300	10.29
91/11/15	A	3	A	PVT	750.00	ALTS & ADD TO OFFICES	4	1536000	1578960	7.42	1567375	2.63
91/11/20	E	8	N	PVT	750.00	SENIORS CENTRE	13	1709547	1735132	29.91	1895540	8.63
91/11/22	E	8	A	PVT	750.00	ALTS & ADDS TO HOUSE	5	666666	688444	16.05	705996	6.01
91/11/25	D	5	N	PUB/1	750.00	HARBOUR OFFICE	5	677776	695856	14.36	734372	4.67
91/11/30	E	2	A	PVT	750.00	EXT TO OFFICES	5	1946649	1998560	5.21	2010751	2.22
TOTAL						243 PROJECTS RECORDED	1925	653896177	1056390738		713561125	1322
MEAN							8	2690986	4348110	18.58	2936548	5.44

DATE	GROSS ESTIMATE	DEVIATION R's	GRDEV % REL	GRABS % DEV	FIXED SUMS ETC	ALLOW %	TENDER NETT	ESTIMATE NETT	NETT DEVIATION	NTDIV % REL	NTABS % DEV	QS
	ESTIMATE DETAILS											
91/08/30	6958000	30000	0.43	0.43	220000	3.16	6708000	6738000	30000	0.45	0.45	9
91/09/20	3633944	-132923	-4.04	4.04	1160800	30.45	2649067	2495944	-153923	-5.81	5.81	17
91/09/10	1538334	146834	10.55	10.55	389467	27.99	1002038	1148872	146834	14.65	14.65	63
91/09/27	1080000	45000	4.35	4.35	211700	20.45	823300	868300	45000	5.47	5.47	17
91/09/30	726000	8000	1.11	1.11	179600	25.05	538400	546400	8000	1.49	1.49	13
91/10/01	1052345	-21655	-2.02	2.02	268000	24.95	806000	784345	-21655	-2.69	2.69	6
91/10/08	9200000	-146200	-1.56	1.56	2200000	23.54	7146200	7000000	-146200	-2.05	2.05	19
91/10/08	5050000	344091	7.31	7.31	1355506	28.79	3350403	3695000	344091	10.27	10.27	1
91/10/09	7580000	60540	0.81	0.81	565450	7.53	6956010	7014550	60540	0.87	0.87	52
91/10/25	1350000	351529	35.21	35.21	0	0.00	998471	1350000	351529	35.21	35.21	14
91/10/31	746000	9132	1.24	1.24	123800	16.80	613068	622200	9132	1.49	1.49	1
91/11/15	1543000	7000	0.46	0.46	250000	16.28	1286000	1293000	7000	0.54	0.54	26
91/11/20	1850000	140453	8.22	8.22	156500	9.15	1553047	1693500	140453	9.04	9.04	21
91/11/22	680994	14328	2.15	2.15	225800	33.87	440866	455194	14328	3.25	3.25	63
91/11/25	567000	-110776	-16.34	16.34	0	0.00	677776	567000	-110776	-16.34	16.34	10
91/11/30	1977700	31051	1.60	1.60	159500	8.19	1787149	1818200	31051	1.74	1.74	5
TOTAL	686336307	32466330			-713079042		-684354184	-710817561	32466330			
MEAN	2824512	133606	4.97	8.56	-2956335	21.32	-2824490	-2958097	-133606	4.32	11.75	

www.ingramcontent.com/pod-product-compliance
Lightning Source LLC
LaVergne TN
LVHW042105190726
843493LV00006B/1359
* 9 7 8 3 3 8 4 2 4 1 4 2 9 *